SOLUTION CHEMISTRY RESEARCH PROGRESS

SOLUTION CHEMISTRY RESEARCH PROGRESS

DARIAN V. BOSTRELLI
EDITOR

Nova Science Publishers, Inc.
New York

For permission to use material from this book please contact us:
Telephone 631-231-7269; Fax 631-231-8175
Web Site: http://www.novapublishers.com

LIBRARY OF CONGRESS CATALOGING-IN-PUBLICATION DATA

Solution chemistry research progress / Darian V. Bostrelli (editor).
p. cm.
ISBN-13: 978-1-60021-993-1 (hardcover)
ISBN-10: 1-60021-993-4 (hardcover)
1. Solution (Chemistry)--Research. 2. Solvents--Thermalmechanical properties. I. Bostrelli, Darian V.
QD544.S65 2007
541'.34--dc22 2007034879

Published by Nova Science Publishers, Inc. ✣ New York

CONTENTS

PREFACE

Solution chemistry deals with liquid solutions in such fields as physical chemistry, chemical physics, molecular biology, statistical mechanics, biochemistry, and biophysics. This book includes experimental investigations of the dielectric, spectroscopic, thermodynamic, transport, or relaxation properties of both electrolytes and nonelectrolytes in liquid solutions. The latest research in the world has been selected, gathered and presented here.

Expert Commentary - Mixtures of water with acetonitrile find much use in several fields of chemistry, such as electrochemistry, chromatography, synthesis, etc. The properties of such mixtures, including thermodynamic, spectroscopic, and transport ones, have been studied extensively. The structure of such mixtures was obtained by both diffraction and computer simulation methods. It is generally agreed now that at intermediate compositions of the mixtures there exists a region of microheterogeneity, although the limits of this region have not been delineated unambiguously. In this region the mixtures act as a single homogeneous phase macroscopically, but clustering occurs at molecular dimensions, the fluctuations in composition being large because self-interactions predominate over mutual ones between the components. As the temperature is decreased from ambient ones this phenomenon augments, till at 271.86 K (–1.3 °C) actual phase separation takes place.

Chapter 1 - Thermomechanic and thermodynamic properties of mixing solvents in the liquid state have proven a powerful tool in elucidating the specific interactions and structural features supporting the liquid structure in multicomponent nonelectrolytic solutions. Mass transport properties (density, viscosity), thermophysical properties (relative permittivity, refractive index), and related thermodynamics can deepen elucidate the solvent – solvent specific interactions and behavioural peculiarities of the complex real systems. In this paper, the authors will primarily deal with the experimental measurements and correlative studies both on temperature and composition, of the above mentioned properties (Y), working with a ternary solvent system containing 1,2-ethanediol (ED), 2-methoxyethanol (ME) and 1,2-dimethoxyethane (DME), at various temperatures in the range $-10 \leq t$ (°C) ≤ 80, and with different compositions covering the whole miscibility field of the selected species. These components are strictly parent species, all being 1,2-ethanedyil-derivatives ($-CH_2-CH_2-$), pertaining to the amphiphile class molecules, with different balancing degree of hydrophobic and hydrophilic properties. Furthermore, the excess properties (Y^E) and deviation functions (ΔY) are largely stressed and examined along the text in order to identify the presence of solvent – cosolvent adducts in these ternary mixtures.

Chapter 2 - Knowledge of the physico-chemical properties of ions in solution is indispensable for understanding their behavior in laboratory, physiological, geological, and industrial environments. Apart from the transference numbers, conductivities and (self) diffusion coefficients that can be measured for individual ions in solution, most other desired properties can only be estimated properly by splitting measured values for electrolytes into the contributions of their constituent ions.

This pertains to such thermodynamic quantities as standard partial molar volumes, heat capacities, and entropies. Also of importance in this context are the standard molar Gibbs energy, enthalpy, and entropy of solvation and of transfer between solvents. The problem has so far found various solutions for obtaining individual ionic properties in selected solvents, foremost water, and to a lesser extent also in binary solvent mixtures.

The desiderata for methods for splitting the electrolyte data are stated, and the merits and disadvantages of existing methods are discussed in detail. In view of these, the requirements of future research in this direction are presented.

Chapter 3 - Molecular recognition of sugars serves as a basic concept to understand biological functions of these compounds and offers a key in developing new types of sugar-based functionalized materials. Since most sugars are isomers which only differ in the stereo arrangements of specific OH-groups, in general the complex formation will be determined by molecular recognition of their structure but not distinctions of chemical reactivity. From this point of view, sterically well-organized macrocyclic compounds such as crown ethers, cyclophanes, cyclodextrines, *etc.* are the most attractive as potential receptors for sugars. This review includes the results reported in literature and own data concerning an ability of mono- and oligosaccharides to form noncovalent intermolecular complexes with natural and synthetic macrocyclic compounds in solutions. Much attention is paid to the effect of stereochemical structure of the sugars on the selective interactions with the macrocyclic ligands. It is an attempt to elucidate the nature of the interactions responsible for molecular recognition of sugars and the role of solvent in this process.

Chapter 4 - To modify the physical-chemical and biological properties of biomolecules and to prolong their biological action, the ability of native and hydroxypropylated α- and β-cyclodextrins to form inclusion complexes with some B-vitamins (pyridoxine, pyridoxal, nicotinamide, nicotinic acid) in aqueous solution at 298.15 K has been studied by means of calorimetry, capillary electrophoresis and ^{1}H NMR spectroscopy. Two thermodynamic approaches were applied to characterize the complexation processes and weak interactions that are not accompanied by complex formation. In the first case, the thermodynamic properties of complexation such as K, $\Delta_c G$, $\Delta_c H$ and $\Delta_c S$ were considered. For interpretation of weak interactions, the enthalpic virial coefficients (h_{xy}) were calculated according to the McMillan-Mayer theory. On the basis of the experimental data, it was found that cyclodextrins exhibit selectivity in the interactions with B-vitamins. The formation of 1:1 enthalpy stabilized complexes was observed in the systems (α-cyclodextrin + nicotinic acid), (hydroxypropyl-α-cyclodextrin + nicotinic acid), (α-cyclodextrin + isonicotinic acid), (α-cyclodextrin + nicotinamide) and (β-cyclodextrin + nicotinamide). The obtained thermodynamic parameters of interactions were discussed in terms of influence of the following factors: the structure of guest molecules; availability of the macrocyclic cavity and its dimensions; the presence of hydroxypropyl substituents and medium pH. Using the ^{1}H

NMR data the possible modes of inclusion of guest molecule into cyclodextrin cavity were proposed.

Chapter 5 - Empirical indices of binary aqueous mixtures of protic and aprotic solvents, representing the electron pair and hydrogen bond exchange ability of these mixtures are correlated to the standard Gibbs energy of adsorption, ΔG^{o}_{ads}, of neutral organophosphorous compounds at the dropping mercury electrode - solution interface.

Correlations are carried out in terms of linear and multilinear empirical relations employing the Kamlet – Taft β_{KT}, α and the Dimroth – Reichardt $E_T(30)$ and Kosower Z parameters, which account for solvent – electrode and solvent – solvent interactions. On the basis of the heterosolvation assumption the effect of each binary mixture on the adsorption equilibrium, expressed by the ΔG^{o}_{ads}, is discussed. It is suggested that the effect of solvent composition on the values of ΔG^{o}_{ads} is predominantly determined by the adsorption of organic cosolvent and the water-organic solvent interactions.

Chapter 6 - Using data available in the literature (mainly from the Timmermans tabulation), the experimental boiling points of electrolyte solutions at constant pressures are expressed as a function of the concentration of the dissolved electrolyte in water. The elevations of boiling points are used in the evaluation of the Pitzer parameters of the activity and osmotic coefficients at the boiling point temperatures for 67 electrolytes of the 1:1, 1:2, 2:1 and 2:2 types.

Chapter 7 - Several ways of obtaining information in ionic processes by the analysis of solubility isotherms have been demonstrated. The computational treatment called «hydration analysis » has been described. The special properties of saturated solutions of hydrates have been demonstrated an discussed. Finally, the information obtainable by the comparative analyses of solubility isotherms in systems with an electrolyte, an organic solvbent and water has been discussed and a formalism to describe the shape of such isotherms has been suggested.

Chapter 8 - Apparent molar volumes, $V_{\Phi,S}$, and viscosity *B*-coefficients for D-(+)-glucose, D-(+)-galactose, D-(+)-xylose, and D-(-)-ribose in aqueous L-serine solutions have been determined respectively from density and viscosity measurements at 298.15 K. The infinite-dilution apparent molar volumes of monosaccharides, $V^{0}_{\Phi,S}$, have been evaluated and used to calculate the corresponding transfer volumes, $\Delta_t V^{0}_{\Phi,S}$, from water to aqueous serine solutions. It is shown that values of transfer volumes and viscosity *B*-coefficients are positive and increase with increasing serine contents. Volumetric interaction parameters, v_{SA}, have been obtained from standard transfer volumes of the monosaccharides. Both values of v_{SA} and viscosity *B*-coefficients are in the order: glucose > galactose; xylose > ribose, which is ascribed to the difference in stereochemistry between *e*–OH and *a*–OH groups. The interactions between monosaccharides and serine are discussed in terms of the structural interaction model and the stereo-structure of monosaccharide molecules. On the other hand, the infrared spectra of the four monosaccharides and 3-O-methyl-D-glycopyranose in deuterium oxide with and without serine were recorded at room temperature. The results confirm further the conclusions from volumetry and viscosity.

In: Solution Chemistry Research Progress
Editor: Darian V. Bostrelli, pp. 1-4

ISBN 978-1-60021-993-1

Expert Commentary

CLUSTERING IN LIQUID MIXTURES OF WATER AND ACETONITRILE

Yizhak Marcus
Hebrew University of Jerusalem, Israel

Mixtures of water with acetonitrile find much use in several fields of chemistry, such as electrochemistry, chromatography, synthesis, etc. The properties of such mixtures, including thermodynamic, spectroscopic, and transport ones, have been studied extensively. The structure of such mixtures was obtained by both diffraction and computer simulation methods [1,2]. It is generally agreed now that at intermediate compositions of the mixtures there exists a region of microheterogeneity, although the limits of this region have not been delineated unambiguously [3,4]. In this region the mixtures act as a single homogeneous phase macroscopically, but clustering occurs at molecular dimensions, the fluctuations in composition being large because self-interactions predominate over mutual ones between the components. As the temperature is decreased from ambient ones this phenomenon augments, till at 271.86 K (−1.3 °C) actual phase separation takes place. [5]

Phase separation in binary liquid mixtures A + B is characterized by the second derivative of the Gibbs energy of mixing, $G^{M}{}_{AB}$, with respect to the composition, e.g. the mole fraction x_B of B, being negative at a given temperature T and pressure P:

$$(\partial^2 G^{M}{}_{AB}/\partial x_B{}^2)_{T,P} < 0 \tag{1}$$

The same, a negative sign, applies also to the chemical potentials, μ_i, activities, a_i, and activity coefficients, f_i, of each component with respect to its mole fraction, x_i. The excess Gibbs energy of mixing, $G^{E}{}_{AB}$, in the case of phase separation, has a negative sign for the expression D:

$$D = RT/\, x_B(1 - x_B) + (\partial^2 G^{E}{}_{AB}/\partial x_B{}^2)_{T,P} < 0 \tag{2}$$

The excess Gibbs energy, as also the chemical potentials, activities and activity coefficients, are generally obtained from the measured vapor pressures and vapor compositions of the liquid mixtures over their entire composition range. Corrections for the association of the vapors are taken into account by means of the second virial coefficients of the latter, but may be neglected when the vapor pressures are small.

There are six reports of the vapor pressures and/or derived excess Gibbs energies of mixtures of acetonitrile with water at ambient pressure (0.1 MPa) at various temperatures where they constitute a single homogeneous phase [6-11]. For some of them, the derived $G^E{}_{AB}$, where A = water and B = acetonitrile, cannot be correct because they lead to $D \leq 0$ over a certain composition range, hence the mixtures ought to split into two phases, which they do not.

The oldest reported partial vapor pressures, by Vierk [6] at 20 and 30 °C disagree with all the subsequently reported data that are in substantial gross agreement with each other, showing only a slight temperature variation. It is unfortunate, therefore, that Matteoli and Lepori [12] employed these data at 30 °C for calculations of the preferential solvation in the mixtures by means of the Kirkwood-Buff approach.

The reported partial pressures and/or excess Gibbs energies of mixing of the other authors [7-11] could be expressed accurately by 4th degree polynomials that served for obtaining the second derivatives needed in eq. (2). The standard errors of the fits were 0.002 to 0.005 kJ mol^{-1} in G^E ranging up to ~1.3 kJ mol^{-1}. The total vapor pressure and vapor compositions reported by Treiner et al. [7] at 298.15 K (25 °C) lead to reported G^E values that are in general agreement with those reported later, but differ in subtle manner from them. In fact these data lead to values of $D < 0$ at $0.30 \leq x_B \leq 0.40$, hence cannot be correct. A similar problem arises with the recent partial vapor pressure data of Nikolova et al. [8] at 293.15 K (20 °C), but over only a very narrow composition region near $x_B = 0.35$, where D is just below zero. Still, if only the partial pressures of the acetonitrile (= B) are employed, disregarding those of water, the first derivative of the resulting chemical potentials, $d\mu_B/dx_B$, are positive throughout the composition range, hence are compatible with the existence of these mixtures as a single homogeneous liquid phase.

The data reported by French [9] at 278.15 K (5 °C) lead to $D < 0$ at $0.30 \leq x_B \leq 0.45$, i.e., ought to have split into two phases, although the system is 6 degrees above the upper consolute temperature. The same problem exists with the data at 288.15 K (15 °C), the region of supposed phase separation being somewhat narrower: $0.33 \leq x_B \leq 0.40$. The cause of this apparent inconsistency is that the actual vapor pressure data were measured at 308.15 K (35 °C) and the data for the lower temperatures were obtained by the use of excess partial molar enthalpies reported by Stokes [13]. Easteal and Woolf [10] used the French data [9] at these two temperatures and arrived at essentially the same results. However, at 298.15 K the French data show slightly negative values of D near $x_B = 0.35$, although Easteal and Woolf, ostensibly using the same data, obtained positive values for this function.

French reported G^E values at 308.15 K (35 °C) and 318.15 K (45 °C) that lead to $D > 0$ over the entire composition, as they should. As stated above, only the 308.15 K data arose from their own vapor pressure measurements, those at the higher temperature were calculated from the partial excess enthalpies [13]. Easteal and Woolf derived their reported data for 313.15 (40 °C) and 323.15 K (50 °C) from the same sources [9,13]. There are, however, G^E data for 323.15 K (50 °C) that can be calculated from the expression and parameters reported by Wilson et al. [11]. None of these G^E values for $T \geq 308.15$ K (≥ 35 °C) suffer from the $D < 0$ problem.

It is noteworthy that the $G^E(x_B)$ obtained from the data in [7-11] are all close together at all the temperatures 278.15 to 323.15 K, and it is only the second derivative that causes some of the results to prove to be incorrect. It is also noteworthy that the errors (i.e., a resulting $D < 0$) are confined to narrower ranges of x_B as T increases.

It is unfortunate, however, that the data for the lowest temperature studied have this error, so that they cannot be used for the estimation of the preferential solvation near but above the consolute temperature by means of the Kirkwood-Buff method as used by Matteoli and Lepori [12]. This becomes possible only with the data at $T \geq 298.15$ K (those of Easteal and Woolf [10], not those of French [9]).

But there arises a new problem, due to values of D being rather small (say < 1 though positive) over some composition range. The value of D occurs in the denominators of the expressions used for the calculation of the Kirkwood-Buff integrals, G_{ij}, hence of the preferential solvation parameters, δx_{ij} [12, 14]:

$$G_{AB} = RT\,[\kappa_T - x_B(1-x_B)V_AV_B/VD] \tag{3}$$

$$G_{AA} = RT\,[\kappa_T + V_B^2/\,x_A^2VD] - V/x_A \tag{4}$$

$$G_{BB} = RT\,[\kappa_T + V_A^2/\,x_B^2VD] - V/x_B \tag{5}$$

$$\delta x_{AA} = x_Ax_B(G_{AA} - G_{BA})/[x_AG_{AA} + x_BG_{AB} + V_{cor\,A} \tag{6}$$

$$\delta x_{AB} = x_Ax_B(G_{AB} - G_{BB})/[x_AG_{AB} + x_BG_{BB} + V_{cor\,B}] \tag{7}$$

where κ_T is the isothermal compressibility of the mixture, V_A and V_B are the partial molar volumes of the components in the mixture, V is the molar volume of the mixture, and V_{cor} is a correlation volume. Small values of the function D lead to diverging values of the integrals, of which relatively small differences are needed to yield the preferential solvation parameters. The onset of $D < 1$ is near $x_B = 0.25$ at all the temperatures studied in [7-11], but the range becomes narrower as T increases, the upper limit being: $x_B \sim 0.65$ at 278 K, 0.60 at 288 K, 0.55 at 298 K, 0.50 at 308 and 313 K, and 0.45 at 318 and 323 K. This composition range agrees substantially with that corresponding to the occurrence of microheterogeneity derived from various measurements. The limit of $D > 1$ for avoiding clustering is arbitrary, but appears to lead to reasonable results for the present system at all the temperatures investigated.

REFERENCES

[1] Takamuku, T.; Tabata, M.; Yamaguchi, A.; Nishimoto, J.; Kumamoto, M.; Wakita, H. Yamaguchi, T. *J. PhDs. Chem. B* 1998, *102*, 8880-8888.

[2] Oldiges, C.; Wittler, K.; Tönsing, T.; Alija, A. *J. PhDs. Chem. B* 2002, *106*, 7147-7154.

[3] Davis, M. I. *Thermochim. Acta* 1984, *77*, 421-437.

[4] Bertie, J. E.; Lan, Zh. *J. PhDs. Chem. B* 1997, *101*, 4111-4119.

[5] Szydłowski, J.; Szykula, M. *Fluid Phase Equil.* 1999, *154*, 79-87.

[6] Vierk, A. L. *Z. Anorg. Chem.* 1950, *261*, 283-296.

[7] Treiner, C.; Tzias, P.; Chemla, M. *J. Chem. Soc., Faraday Trans. 1* 1967, *72*, 2007-2015.

[8] Nikolova, P. V.; Duff, S. J. B.; Westh, J.; Haynes, C. A.; Kasahara, Y.; Nishikawa, K.; Koga, Y. *Can. J. Chem.* 2000, *78*, 1553-1560.

[9] French, H. T. *J. Chem. Thermodyn.* 1987, *19*, 1155-1161.

[10] Easteal, A. J.; Woolf, L. A. *J. Chem. Thermodyn.* 1988, *20*, 701-706.

[11] Wilson, S. R.; Patel, R. B.; Abbott, M. M.; Van Ness, H. C. *J. Chem. Eng. Data* 1979, *24*, 130-132.

[12] Matteoli, E.; Lepori, L. *J. Chem. Phys.* 1984, *80*, 2856-2863.

[13] Stokes, R. H. *J. Chem. Thermodyn.* 1987, *19*, 977-983.

[14] Marcus, Y. *J. Solution Chem.* 2006, *35*, 251-277.

In: Solution Chemistry Research Progress
Editor: Darian V. Bostrelli, pp. 5-49 ISBN 978-1-60021-993-1

Chapter 1

MODELLING OF EXPERIMENTAL THERMOPHYSICAL DATA BY MIXING OF A TERNARY SOLVENT SYSTEM

***Giorgia Foca*[1], *Andrea Marchetti*[2],**
***Lorenzo Tassi*[2] *and Alessandro Ulrici*[1]**

[1] Dipartimento di Scienze Agrarie ed Alimentari, Univ. di Modena e Reggio Emilia,
Via Amendola 2, 42100 Reggio Emilia, Italy

[2] Dipartimento di Chimica, Univ. di Modena e Reggio Emilia,
Via G. Campi 183, 41100 Modena, Italy

ABSTRACT

Thermomechanic and thermodynamic properties of mixing solvents in the liquid state have proven a powerful tool in elucidating the specific interactions and structural features supporting the liquid structure in multicomponent nonelectrolytic solutions. Mass transport properties (density, viscosity), thermophysical properties (relative permittivity, refractive index), and related thermodynamics can deepen elucidate the solvent – solvent specific interactions and behavioural peculiarities of the complex real systems. In this paper, we will primarily deal with the experimental measurements and correlative studies both on temperature and composition, of the above mentioned properties (Y), working with a ternary solvent system containing 1,2-ethanediol (ED), 2-methoxyethanol (ME) and 1,2-dimethoxyethane (DME), at various temperatures in the range $-10 \leq t$ (°C) ≤ 80, and with different compositions covering the whole miscibility field of the selected species. These components are strictly parent species, all being 1,2-ethanedyil-derivatives ($-CH_2-CH_2-$), pertaining to the amphiphile class molecules, with different balancing degree of hydrophobic and hydrophilic properties. Furthermore, the excess properties (Y^E) and deviation functions (ΔY) are largely stressed and examined along the text in order to identify the presence of solvent – cosolvent adducts in these ternary mixtures.

INTRODUCTION

The study of thermomechanic and thermodynamic properties of multicomponent solvent mixtures contributes to deepen the understanding of the structuring features of different liquids with different functional groups, which in turn, develop the more strong intermolecular specific interactions sustaining the liquid state and conditioning the macroscopic behaviour of such solvent systems. These information are very useful from both the point of view of theoretical and applied chemistry, because of mixed solvents are largely utilised in the design of industrial processes, engineering purposes, for analytical applications, as far as in the development of theories for the liquid state and predictive methods.

Since a long time our research group have been engaged along these research activities. Several properties of a variety of liquid mixtures were deeply investigated and documented in our studies. For this work, we have chosen three components of parent basic structure, all being 1,2-ethanedyil ($-CH_2-CH_2-$) - derivatives, namely 1,2-ethanediol (ED) and its mono- and di-methyl ethers, such as 2-methoxyethanol (ME) and 1,2-dimethoxyethane (DME), respectively.

ED is a very interesting chemical subject for the study of thermophysical and related thermodynamic properties because of its intrinsic and constitutive remarkable characteristics. This solvent is the simplest homologue of the diols series. It shows a high thermal capacity and molecular stability in relation to a very high surface tension. For these pourposes, ED is widely used as thermoregulator fluid, as a controlling agent of density and viscosity reaction baths for several industrial processes and syntheses as well [1]. Furthermore, for its unusual viscokinetic characteristics (also observed for all other common diols), ED shows a behaviour that lies between Newtonian liquids (at high temperatures, $t > 50$ °C), *quasi-non*-Newtonian (or *pseudo*-rheological) fluids (at ambient - intermediate temperatures), and true rheological systems (such as pastes, muds, suspensions, high polymers, ecc.) at low temperatures ($t < 0$ °C). ED is an excellent solvent for a variety of organic and inorganic species. It is characterised by an amphiprotic nature, potentially acidic medium with a high autoprotolysis constant ($pK_{autoprot} = 15.8$ at 25 °C), and parity of HBDA (Hydrogen - Bonding - Donor - Acceptor) properties (two active sites, acting both as donor and acceptor sites). All these characteristics are also associated to a high value of the dipole moment ($\mu_{ED} = 2.28$ D; 1 D = 3.3356×10^{-30} C ·m) and to a moderately high relative permittivity ($\varepsilon = 37.7$ at 25 °C), classifying this solvent as a dissociating medium. For these reasons among others, it has been largely used in our laboratories for some previous electroanalytical studies [2 - 4].

ME is a hydroxyether, showing *quasi*-aprotic character with a low autoprotolysis constant ($pK_{autoprot} = 20.5$ at 20 °C) and a low relative permittivity ($\varepsilon = 16.94$ at 25 °C). Further, eventhough it is characterised by a high dipole moment ($\mu_{ME} = 2.36$ D), ME can be classified as a scarcely dissociating solvent either for ionogen or ionophore solutes [5,6]. Due to its constitutive features, ME shows unparity of HBDA properties (one donor and two acceptor sites).

DME has attracted much attention in the past years, and particularly during the last two decades, owing to its promising characteristics as non-aqueous solvent for lithium power batteries and others potential applications in the field of supercapacitors and superconductors [7,8]. It is an aprotic protophilic medium, with a small dipole moment ($\mu_{DME} = 1.17$ D) and low relative permittivity ($\varepsilon = 7.54$ at 25 °C). For these reasons, DME can be classified as a not dissociating solvent whose solvating power is manifested by acting as an HBA (Hydrogen - Bonding - Acceptor) species, or as a stabilizing agent *via* CT (Charge Transfer) complexes formation by active sites towards both nonelectrolytic molecules and ion-pairs solutes.

In summary, the simplest amphiphilic species ED, ME and DME show an increasing structural complexity, obtained by the progressive substitution of the hydroxylic –H atoms with the corresponding voluminous methylic ($-CH_3$) groups. As a consequence, these different functional groups and their distribution on the skeleton of the molecules, obviously accomplish for the different physico-chemical properties of the components [9]. This way, being these chemicals perfectly mixable one each other at any ratio and at any experimental condition, their ternary mixtures become an interesting medium where it is possible to vary the reciprocal compositions of the system to obtain the performing features (in terms of HBDA characteristics) by "dosing" the number of –OH and $-OCH_3$ groups per unit volume of the liquid system.

In other words, from these studies, we wish to realize a "Solvent Engineering Project", that is a virtual synthesis of the desired properties obtained by mixing real components, in order to modulate the micro- and macroscopic behaviour of the solvent systems.

EXPERIMENTAL

Materials

The solvents used in this work, ED, ME and DME (containing water < 0.10 %, < 0.05 % and < 0.10 % by mass, respectively, as found by Karl-Fischer titrations) were high purity grade reagents from Carlo Erba (Milan). ED and ME were partially degassed by ultrasonics and stored over molecular sieves (type 3Å from Fluka) for many days before their use. DME was further purified by double fractional distillation over $LiAlH_4$ to eliminate traces of acids and peroxides, and to reduce as much as possible, the total amount of water, keeping the middle fraction only (bp t = 83 °C) for the measurements. The final purity of the solvents was checked by gas chromatography (mass fraction = 99.5 % for ED and 99.7 % for both ME and DME species), confirming the absence of other organic compounds. Experimental densities, viscosities, static relative permittivity and refractive index for the pure components well agree with literature values [10-13].

Apparatus and Procedures

Mixtures were prepared by weighing an appropriate volume of each solvent (Mettler PM 480 Δ-range analytical balance, precision 10^{-4} g). Each mixture was prepared in an airtight bottle to avoid preferential evaporation and in a dry box to avoid atmospheric contamination. The estimated accuracy for each mole fraction (x_i) is estimated to be less than 1.5×10^{-4}.

Densities of the components and their mixtures were measured by using a vibrating tube density meter (Anton Paar, Model DMA 60), equipped with a density measuring cell (DMA 602) with a sensitivity of $\pm 1 \times 10^{-6}$ g cm^{-3}, while the estimated accuracy (95% confidence interval) on density values (ρ) was $\pm 3 \times 10^{-6}$ g cm^{-3} (for more experimental details see ref. [14]).

Kinematic viscosities were measured by a Schott–Geräte AVS 400 instrument, equipped with a series of Ubbelhode viscosimeters covering the $0.3 \leq \nu/\text{cSt} \leq 300$ range (1 cSt = 10^{-6} m^2 s^{-1}). In all the determinations the kinetic energy correction has been taken into account according to the method suggested by Hagenbach [15]. Experiments were replicated at least five times for each composition and temperature, and the mean values of the results were considered. The experimental reproducibility of the flow time at each temperature and composition was ±0.05%, leading to a standard deviation in the viscosity evaluation ranging from 1.0×10^{-4} to 1.0×10^{-1} cSt, for the lowest and the highest measured values, respectively. As a consequence, the estimated accuracy (95% confidence level) ranges from $\pm 3 \times 10^{-4}$ cSt to $\pm 3 \times 10^{-1}$ cSt [15].

The static relative permittivities were measured by the heterodyne beat method at a frequency of 2 MHz with a WTW – DM01 Dipolmeter, equipped with two thermostated stainless steel measuring cells type MFL 2 ($7 < \varepsilon < 21$) and MFL 3 ($21 < \varepsilon < 90$). The thermostated measuring cell was embedded in a thick polyurethane protective jacket, and the temperature control was provided by a Lauda K2R thermostatic bath maintained to ±0.02 K. The temperature was checked by a thermoresistance Pt 100 (Tersid, Milan) immersed in the measuring cell. This device shows a slight quadratic dependence of the resistance with temperature [$R(T) = R_{273.15}(1 + AT + BT^2)$] in the investigated thermal range, and with tolerance of ±0.01 K. The resistance values were collected using a Wayne Kerr 6425 PCA (Precision Component Analyzer). The experiments were generally performed for at least 10 replicate runs (without refilling the cell) for each composition and temperature, and the results were averaged. Reproducibility of measurements [standard deviations $\sigma(\varepsilon)$] was approximately ± 0.2%, and the estimated accuracy (at 95% confidence level) was in the range $\pm 2 \cdot 10^{-2}$ (for the largest ε measured) and $\pm 5 \cdot 10^{-3}$ (for the smallest ε measured). Further details about the apparatus and procedures for the static dielectric constant measurements have been described elsewhere [16].

The refractive indeces n_D (Na - D line, at $\lambda = 589$ nm) were measured by an automatic refractometer GPR 11-37X Index Instruments, with a resolution of 1×10^{-5} and an accuracy of $\pm 5 \times 10^{-5}$. The instrument operates in the temperature range $0 \leq t\ /\ °\text{C} \leq 70$. The measuring cell, operating in continuos mode, was calibrated with pure water (from a Milli Q-Plus apparatus, Millipore) at each temperature. The measurements of the n_D values, for each sample and at each temperature, were repeated at least five times by refilling the cell. For the measurement of n_D a special care was adopted, for each sample and at each temperature, and the reading was taken whenever its value in pre- and post-calibration cell with water differed by less than ±0.00005. The reproducibility of the measurements, expressed as standard deviation $\sigma(n_D)$, was approximately equal to 6×10^{-5}, and the accuracy was $\pm 7 \times 10^{-5}$ for a 95% confidence level. The temperature was maintained constant within ± 0.01 °C by an Haake F3C thermostatic bath [17].

RESULTS AND DISCUSSION

Density, and Volumic Properties

The results reported in this section concerned with the volumic properties of the {ED(1) + ME(2) + DME(3)} ternary system. The experimental density measurements were collected in the temperature range $-10 \leq t\ /\ °C \leq 80$, with thermal scansion of 5 °C, working with 12 three-component mixtures covering the whole miscibility field, as expressed by the relation $0 < x_1, x_2, x_3 < 1$. Following this procedure, we capitalize our previous results obtained by studying the volumetric behaviour of the three binary subsystems {ED(1) + ME(2)} [18], {ED(1) + DME(3)} [19], and {ME(2) + DME(3)} [20].

This way, the ternary solvent system covers, on the whole, the three pure species, 27 binaries containing {ED, ME, DME}, and 12 {ED + ME + DME}, mixtures (total: 19 isothermal sets, each of them composed of 42 data points). To reduce the total number of tables, the paper reports only the essential material in order to make more fluent the reading, and exploiting the best visual impact of the figures rather than other choices.

Experimental densities are listed in Table 1, along with the ternary compositions expressed as mole fractions. These data have been processed by applying different relationships describing the functional dependence on temperature, $\rho = \rho\ (t)$, and composition, $\rho = \rho\ (x_i, x_j)$, with the aim to obtain the best generalization of the most widespread and currently applied models [21].

Even though the knowledge of density data for liquid systems is of fundamental importance in many fields, it has been stressed in many reports that industrial design needs the knowledge of another useful related parameter, namely the excess molar volume (V^E). Furthermore, the excess properties are analyzed because of their importance for inferring which type of interactions predominates in liquid mixtures. Starting from the experimental density data, for a generic n-components solvent mixture, the excess molar volume is calculated with the equation

$$V^E = \frac{\sum_{i=1}^{n} x_i M_i}{\rho} - \sum_{i=1}^{n} \frac{x_i M_i}{\rho_i} \tag{1}$$

where: M_i are the molar masses ($M_1 = 62.070$; $M_2 = 76.096$; $M_3 = 90.120$ g mol^{-1} for ED, ME, DME, respectively), and the other symbols have their usual meaning.

Some indications of the complexity of the ternary liquid system as a result of the variation in ternary composition is given by the $V^E = V^E(x_i)$ surface in Figure 1.a, related to the results at 25 °C. This figure has been obtained by plotting the smoothing equation [22]:

$$V^E\ (Y^E \text{ or } \Delta Y) = d_1 x_1 x_2 + d_2 x_2 x_3 + d_3 x_1 x_3 + d_4 x_1 x_2 (x_2 - x_1) + d_5 x_2 x_3 (x_3 - x_2) + d_6 x_1 x_3 (x_1 - x_3) + d_7 x_1 x_2 x_3 \tag{2}$$

Table 1. Ternary composition (x_i) and experimental densities (ρ/g cm^{-3}) for the {ED (1) + ME (2) + DME (3)} solvent system at various temperatures

x_1	1	0	0	0.3192	0.4727	0.1516	0.1871	0.0770	0.1323	0.3742	0.7191	0.2987	0.8588	0.1237	0.6022
x_2	0	1	0	0.5722	0.5143	0.5299	0.3028	0.6925	0.7890	0.3964	0.1957	0.1963	0.0738	0.0924	0.1539
x_3	0	0	1	0.1086	0.0131	0.3185	0.5101	0.2305	0.0788	0.2294	0.0852	0.5050	0.0675	0.7839	0.2439
t/°C															
-10	1.13363_4	0.99016_9	0.89785_3	1.03138_2	1.05693_7	0.98589_3	0.96633_1	0.98340_1	1.00512_9	1.02638_7	1.08236_7	0.97885_6	1.10346_4	0.925135	1.049686
-5	1.13045_2	0.98637_7	0.89276_8	1.02669_0	1.05240_3	0.98111_7	0.96161_6	0.97898_2	1.00088_6	1.02144_9	1.07913_8	0.97423_9	1.10056_7	0.920479	1.045765
0	1.12706_9	0.98234_5	0.88772_3	1.02197_1	1.04792_1	0.97626_8	0.95682_1	0.97438_2	0.99649_3	1.01662_4	1.07577_5	0.96954_1	1.09740_3	0.915732	1.041778
5	1.12366_4	0.97812_8	0.88258_1	1.01727_9	1.04351_0	0.97138_8	0.95196_9	0.96965_2	0.99198_5	1.01187_8	1.07222_4	0.96479_4	1.09402_3	0.910833	1.037707
10	1.12029_6	0.97374_1	0.87735_4	1.01255_8	1.03912_4	0.96648_4	0.94707_6	0.96481_8	0.98738_4	1.00715_9	1.06853_1	0.95999_7	1.09045_3	0.905879	1.033563
15	1.11690_6	0.96925_9	0.87212_5	1.00784_2	1.03475_7	0.96148_0	0.94210_7	0.95982_6	0.98267_5	1.00249_1	1.06469_9	0.95516_4	1.08678_7	0.900820	1.029346
20	1.11344_0	0.96485_1	0.86687_5	1.00314_6	1.03038_3	0.95648_2	0.93712_8	0.95477_5	0.97787_2	0.99781_1	1.06079_8	0.95027_9	1.08304_7	0.895737	1.025063
25	1.10991_3	0.96028_8	0.86150_6	0.99840_7	1.02600_1	0.95140_6	0.93213_9	0.94962_3	0.97299_8	0.99312_5	1.05682_2	0.94539_3	1.07924_1	0.890629	1.020729
30	1.10651_2	0.95574_9	0.85609_7	0.99364_3	1.02160_5	0.94631_5	0.92705_5	0.94439_4	0.96808_5	0.98841_1	1.05280_1	0.94046_5	1.07540_1	0.885474	1.016338
35	1.10308_9	0.95108_8	0.85068_0	0.98884_0	1.01716_2	0.94115_6	0.92198_7	0.93907_5	0.96312_6	0.98366_6	1.04871_0	0.93551_7	1.07157_6	0.880333	1.011897
40	1.09974_7	0.94629_2	0.84512_4	0.98407_2	1.01272_7	0.93600_8	0.91689_8	0.93376_5	0.95813_4	0.97886_9	1.04456_8	0.93054_9	1.06772_0	0.875162	1.007404
45	1.09631_2	0.94162_1	0.83957_5	0.97924_7	1.00827_9	0.93079_9	0.91176_2	0.92840_2	0.95315_8	0.97403_2	1.04037_7	0.92554_6	1.06385_9	0.869976	1.002868
50	1.09294_6	0.93687_8	0.83395_6	0.97437_8	1.00382_4	0.92554_6	0.90660_1	0.92301_3	0.94815_1	0.96913_8	1.03615_5	0.92053_6	1.05999_8	0.864739	0.998281
55	1.08947_1	0.93213_3	0.82824_7	0.96951_0	0.99935_5	0.92026_1	0.90140_3	0.91764_6	0.94312_4	0.96420_3	1.03187_1	0.91549_8	1.05611_9	0.859545	0.993626
60	1.08601_1	0.92738_7	0.82246_8	0.96461_2	0.99490_1	0.91497_1	0.89622_3	0.91227_5	0.93811_6	0.95923_7	1.02756_9	0.91041_5	1.05218_1	0.854257	0.988911
65	1.08263_4	0.92271_4	0.81667_6	0.95970_7	0.99046_7	0.90967_0	0.89104_7	0.90693_2	0.93309_7	0.95424_4	1.02325_9	0.90530_3	1.04824_8	0.848960	0.984167
70	1.07900_3	0.91781_1	0.81083_0	0.95474_6	0.98607_2	0.90433_3	0.88585_9	0.90160_9	0.92809_6	0.94921_2	1.01885_5	0.90010_5	1.04418_0	0.843633	0.979318
75	1.07536_9	0.91308_7	0.80494_7	0.94974_8	0.98172_3	0.89898_7	0.88066_3	0.89631_2	0.92308_3	0.94418_2	1.01440_5	0.89483_7	1.04001_6	0.838237	0.974379
80	1.07169_9	0.90814_4	0.79906_4	0.94468_9	0.97743_9	0.89359_7	0.87543_0	0.89106_5	0.91806_8	0.93914_9	1.00987_6	0.88943_5	1.03572_9	0.832714	0.969314

which has been applied to each isothermal set of 42 values of V^E, and where the coefficients d_k are empirical fitting parameters.

This equation formally derives from the well-known Redlich-Kister one (valid for binary mixtures):

$$V_{ij}^E = x_i x_j \sum_0^k c_k (x_j - x_i)^k \tag{3}$$

This variable-degree function, when written in the simplest expansion with the polynomial truncated at the first power term and adapted to ternary systems, takes the form of equation (2). Furthermore, it represents the V^E quantity of the ternary system as a sum of contributions due to the three binary subsystems

$$V^E_{123} = V^E_{12} + V^E_{23} + V^E_{13} + \delta_{123} \tag{4}$$

and an overall ternary mixture effect (δ_{123}) explicitly contained into the d_7 term of equation (2).

Looking at Figure 1, it is possible to observes that the trend is generally negative, the $V^E = V^E(x_i)$ function taking values in the $0 \geq V^E$ / cm^3 mol^{-1} ≥ -1.55 range at 25 °C. Furthermore, Figure 1.b shows that along the three binary axes it is present a sharp minimum centred at $x_1 \cong 0.5$ for {ED(1) + ME(2)} (stoichiometric ratio = 1ED: 1ME) mixtures, at $x_2 \cong 0.65$ for {ME(2) + DME(3)} (2ME: 1DME), and at $x_3 \cong 0.35$ for the {ED(1) + DME(3)} (2ED: 1DME) binaries. The mixing process of the selected components induces the volume contraction ($V^E < 0$) and, as far as binaries are concerned, it is more sensitive in the sequence {ED(1) + DME(3)} > {ED(1) + ME(2)} > {ME(2) + DME(3)}.

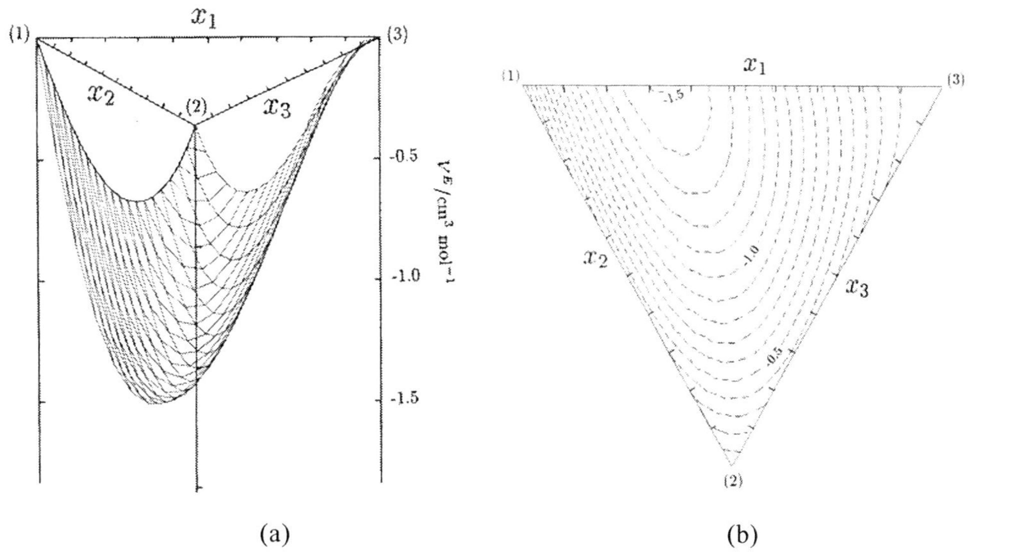

Figure 1. (Continued)

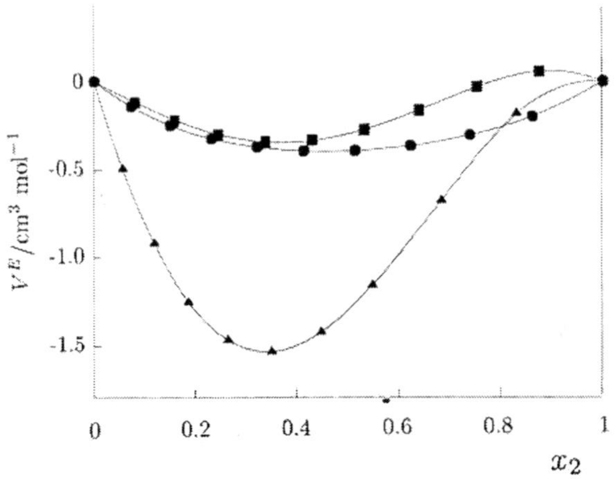

(c)

Figure 1. (*a*) Three-dimensional representation of the experimental data of V^E at 25 °C for the ternary mixtures {ED(1) + ME(2) + DME(3)}. Full-line reticulates represent the surfaces generated by Eq (2). (*b*) Computer generated contour diagram showing lines of constant V^E on a liquid – mole fraction (x_i) grid. (*c*) Profiles of the V^E with changing x_i of binary subsystems: ● {ED(1) + ME(2)}; ▲ {ED(1) + DME(2)}; ■ {ME(1) + DME(2)}.

Several and sometimes opposite effects may contribute to the V^E data. In the present case, we can single out some different contributions that will be discussed in order to give a rationale about the non-ideal behaviour of this solvent system.

i) At first, one may take into account the breaking of the structure of the ordered pure liquids by mixing one each others. ED is certainly one of the most three-dimensionally ordered liquids (at low temperatures), while a linear – bidimensional ordering is preferred at medium temperatures, and such order decreases progressively as T increases [23]. The same consideration is almost valid for ME species too, even if some very consistent differences are present with respect to pure ED. These considerations well agree with the literature suggestions about the molecular dynamics of the selected components when investigated by NMR techniques, NIR spectra recorded at various temperatures and dielectric relaxation spectroscopy, among other methodologies [24-26]. From these studies, it has been demonstrated that all these molecules can exist in different rotameric conformations (namely *anti* and *gauche*) with respect to the two substituting groups in the alkyl chain (-OH, and –OCH_3). In the case of ED and ME, in the *gauche* conformation a strong intramolecular hydrogen bond is present between the two groups (whose energy is estimated to be about 24 kJ mol^{-1} at ordinary conditions [27]).

This skeletal arrangement resembles the molecules to a pseudo-cyclic form, having density and entropy values higher than the more or less linear *anti* conformers. So all molecules should be engaged in participating to the equilibrium

$$\text{non-bonded conformation } (anti) \rightleftarrows \text{ bonded conformation } (gauche)$$

which appears shifted to the right, being the thermodynamic constant K = 10.3 at 20°C for ME [27].

Obviously the molecular arrangement is quite different for the DME units, because of the total absence of HBD characters, and where the *anti* conformer is more largely stabilized by the presence of steric hindrances due to the two voluminous $-OCH_3$ substituent groups [26]. In the case of the mixtures of the present study, one of the effects that may contribute to V^E is the disruption of such orientational correlations and homocooperative interconnections of ED and ME, in particular. In a reliable abstraction, this should correspond to a volumic expansion, and a positive contribution to V^E should be expected.

ii) The second important factor that may be invoked to modulate the V^E quantity is the unfavourable interactions between groups, i.e. between the hydrophobic and apolar groups ($-CH_3$ and $-CH_2-$ groups of alkyl chains), and the polar groups of the three components (-OH and -O- atoms), thus producing valuable microheterogeneity at molecular level [28].

Generally, a positive contribution to V^E from this structural factor is expected if the existence of unfavourable interactions between groups is ascertained, as in the present case, and *vice versa*. Nevertheless, in this kind of analysis, it is necessary to consider all types of interactions between groups. For these three species, it is possible to identify favourable interactions, such as strong hydrogen bonds, between the hydroxyl groups of ED and ME and the ether-oxygen atoms acting as hydrogen acceptor sites. On this basis, a volume contraction should be expected on mixing the pure components when favourable interactions between groups occur. For ternaries studied here, both types of interactions between groups occur and, as a consequence, their final effect should be almost balanced.

iii) Another effect which is expected to act on the V^E property, and which should give a negative contribute to the excess quantity, is the difference in molecular size (geometrical effects) between the components in the mixed system. In this case (V_{ED} = 55.928 $cm^3\ mol^{-1}$, V_{ME} = 79.243 $cm^3\ mol^{-1}$, V_{DME} = 104.606 $cm^3\ mol^{-1}$, at 25 °C), the unlike species have quite different molecular volume; therefore this effect should be very important. In fact the difference in the free volumes between different species could facilitate the compenetration of one component into the other, and as the difference of the free volumes of the pure species increases, the more negative should be the contribution to V^E.

After these considerations, we may assert that in the {ED(1) + ME(2) + DME(3)} ternary mixtures all the effects discussed above should be taken into account and may contribute to V^E, although the structural ones ii) and iii) should prevail, and could explain why the V^E is always negative in the temperature range $-10 \leq t\ /\ °C \leq 80$. Therefore, the attractive interactions (which are generally responsible for structure-making effects [29]) seem to be the driving forces in the liquid structure of these mixed nonelectrolytic solutions, with respect to repulsive phenomena due to sterical hindrances and unfavourable interactions between polar and apolar groups (which, in turn, provide structure-breaking effects).

To gain further insight about the volumic behaviour of these liquid mixtures, an examination of the partial derivatives of the excess properties seems to be of particular interest. The partial molar excess volume $\overline{V}_i^E$ of the *i*-th species in a multicomponent mixture can be determined from the excess molar volume by the equation [30]:

$$\overline{V}_i^E = V^E + (1 - x_i)\frac{\partial V^E}{\partial x_i} \tag{5}$$

from which the partial molar volume $\overline{V_i}$ can be evaluated as follows:

$$\overline{V_i} = V_i + \overline{V}_i^E \qquad (6)$$

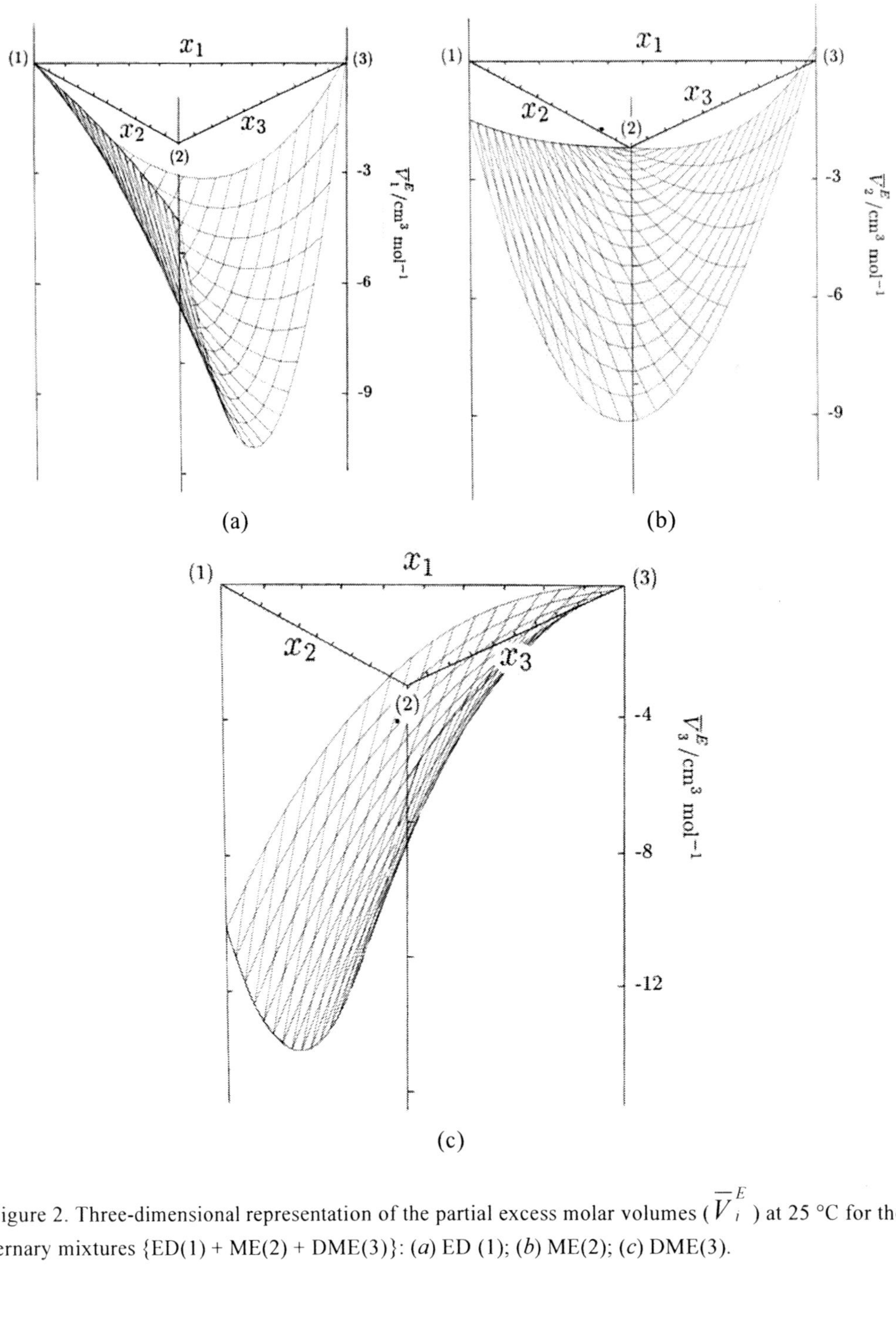

Figure 2. Three-dimensional representation of the partial excess molar volumes ($\overline{V}_i^E$) at 25 °C for the ternary mixtures {ED(1) + ME(2) + DME(3)}: *(a)* ED (1); *(b)* ME(2); *(c)* DME(3).

where V_i is the molar volume of the i-th pure component at each temperature. Considering the relationship (2), and making use of the d_k fitting coefficients, the quantity $\partial V^E/\partial x_i$ is calculated at each temperature and then, using equations (5) and (6), the $\overline{V}_i^E$ and $\overline{V}_i$ values are determined.

As far as $\overline{V}_i^E$ is concerned, the trend for the three species at 25 °C is reported in Figure 2. The greatest evidence for these functions $\overline{V}_i^E = \overline{V}_i^E \{x_1, x_2, x_3\}$ consists in the generally negative trend for all the components.

In particular, the negative values of this property can be indicative of:

- the presence of strong attractive interactions *via* hydrogen bonding, probably followed by a further compressive contribution due to the intermolecular electrostatic interactions from dipolar groups;
- decreasing of structural effects for the pure components by mixing ;
- reciprocal accommodation of each species into molecular cavities of each other component (interstitial accommodation sustained by geometrical effects and steric hindrances).

For a best comprehension, Table 2 summarizes the extrapolated $\overline{V}_i^E$ values at infinite dilution for each species. It appears, from the reported data, that DME is the component with the greatest (negative) deviations of $\overline{V}_i^E$, and this fact perfectly agrees with the low density values, the great molar volumes and the molecular dynamic of this species. Furthermore, the strong volumic contraction for this component (DME) when present at infinite dilution in ED ($\overline{V}_i^E \cong -10.37$ cm^3·mol^{-1}, in turn $\cong$ -10 % of the molar volume), probably should be due to a very consistent compression work exerted by ED molecules, whose HBD ability should be so efficient up to reduce as much as possible the free rotational degrees of -OCH_3 groups.

However, as an interesting detail, we mention that $\overline{V}_i^E$ becomes slightly > 0, and in a very restricted domain, for ED and ME at infinite dilution in the DME cosolvent. A possible rationale for this observation, can be given as follows: the intramolecular hydrogen bonds, characteristic for the *gauche* rotamers, probably, completely break down in extreme dilute region (near pure DME). This fact provides a volumic expansion of the components ED and ME, which definitely rearrange themselves as *anti* conformers (low density low entropy forms), with the pronounced elongation of the skeletal structure. We remark that the volumic expansion ($\overline{V}_i^E > 0$) is not at all considered a reliable index for a slackening of specific intermolecular interactions, because the presence of strong intermolecular hydrogen bonding can still sustain the liquid structure and can still guarantee the formation of heteroaggregated clusters, even if in presence of geometric effects and steric hindrances.

Finally, it is possible to observe that the property $\overline{V}_i^E$ takes the same limiting value at infinite dilution in the case of ED and ME, reciprocally, even if in presence of very different structural features for the two components. This particularity probably confirms that the elongation of the molecular skeleton by rupting the intramolecular hydrogen bonding provides the same effects on the expansion of the molecular volume for these species, and independently whichever the substituent groups (-OH or $-OCH_3$).

As fruitful exercise, it is possible to evaluate the HBDA character of a generic mixture containing ED, ME and DME, this feature being quantifiable and expressed as a sum of contributions due to each component:

$$HBD = 2\,x_1 + 1\,x_2 + 0\,x_3 \qquad 0 \leq HBD\text{ / sites per mole of mixture} \leq 2$$
$$HBA = 2\,x_1 + 2\,x_2 + 2\,x_3 = 2$$

Table 2. Limiting values of partial molar excess volumes, $\overline{V}_i^E$ / $cm^3 \cdot mol^{-1}$, obtained for each component at infinite dilution into each other at 25 °C

Infinite dilution →	ED (1)	ME (2)	DME (3)
i-th species ↓			
ED (1)	---	-1.80	0.28
ME (2)	-1.80	---	0.38
DME (3)	-10.37	-2.80	---

Table 3. Ternary composition (x_i) and experimental kinematic viscosity values (ν / cSt *) for the {ED(1) + ME(2) + DME(3)} solvent system

x_1	1	0	0	0.3192	0.4727	0.1516	0.1871	0.0770	0.1323	0.3742	0.7191	0.2987	0.8588	0.1237	0.6022
x_2	0	1	0	0.5722	0.5143	0.5299	0.3028	0.6925	0.7890	0.3964	0.1957	0.1963	0.0738	0.0924	0.1539
x_3	0	0	1	0.1086	0.0131	0.3185	0.5101	0.2305	0.0788	0.2294	0.0852	0.5050	0.0675	0.7839	0.2439
t / °C															
-10	98.59	3.594	0.7002	7.079	14.47	2.786	1.956	2.839	4.524	5.969	25.92	2.468	45.32	1.042	10.92
-5	71.09	3.133	0.6557	5.914	11.61	2.447	1.754	2.494	3.891	5.001	19.89	2.180	33.78	0.9684	8.788
0	52.42	2.763	0.6220	5.002	9.489	2.166	1.580	2.210	3.391	4.230	15.51	1.939	25.66	0.9070	7.159
5	39.54	2.446	0.5898	4.285	7.878	1.931	1.430	1.971	2.957	3.666	12.65	1.740	20.31	0.8409	6.076
10	30.44	2.182	0.5605	3.699	6.620	1.730	1.298	1.767	2.603	3.185	10.36	1.567	16.25	0.7844	5.162
15	23.89	1.959	0.5338	3.231	5.623	1.566	1.192	1.599	2.316	2.799	8.595	1.425	13.16	0.7371	4.433
20	19.04	1.769	0.5095	2.847	4.824	1.425	1.097	1.452	2.060	2.502	7.344	1.305	10.98	0.6876	3.915
25	15.44	1.607	0.4859	2.515	4.165	1.298	1.011	1.325	1.860	2.206	6.132	1.193	9.032	0.6529	3.363
30	12.67	1.469	0.4626	2.244	3.633	1.192	0.9414	1.216	1.687	1.975	5.221	1.101	7.557	0.6195	2.944
35	10.49	1.341	0.4417	2.012	3.194	1.097	0.8737	1.118	1.528	1.784	4.544	1.017	6.460	0.5835	2.629
40	8.806	1.232	0.4235	1.813	2.824	1.014	0.8153	1.034	1.397	1.610	3.931	0.9439	5.512	0.5558	2.329
45	7.482	1.137	0.4052	1.649	2.521	0.9425	0.7648	0.9603	1.283	1.472	3.463	0.8810	4.787	0.5281	2.101
50	6.434	1.052	0.3887	1.500	2.256	0.8769	0.7177	0.8938	1.182	1.342	3.050	0.8231	4.170	0.5043	1.889
55	5.532	0.9781	0.3735	1.376	2.036	0.8199	0.6774	0.8353	1.094	1.237	2.716	0.7749	3.665	0.4827	1.721
60	4.777	0.9122	0.3592	1.265	1.848	0.7677	0.6379	0.7825	1.014	1.144	2.446	0.7282	3.250	0.4599	1.580
65	4.192	0.8581	0.3456	1.168	1.677	0.7231	0.6067	0.7357	0.9465	1.062	2.196	0.6907	2.888	0.4426	1.447
70	3.688	0.7992	0.3332	1.090	1.538	0.6867	0.5805	0.6949	0.8838	1.000	2.006	0.6598	2.601	0.4250	1.352
75	3.279	0.7504	0.3210	1.009	1.411	0.6460	0.5489	0.6557	0.8271	0.9262	1.817	0.6214	2.339	0.4081	1.241
80	2.909	0.7076	0.3093	0.9463	1.302	0.6171	0.5282	0.6224	0.7743	0.8819	1.692	0.5973	2.144	0.3908	1.179

* 1 cSt = 10^{-6} m^2 s^{-1}.

The parameter that measures HBA, for these ternary mixtures, is a constant. In particular, HBA has the same value (HBA = 2 sites per mole of mixture) whichever the relevant composition of the solvent system. On the contrary, HBD capacity varies continuously between the limits 0 ÷ 2 (from pure DME to pure ED); this way, it is possible to modulate the intrinsic features also in terms of HBD and HBA ability by combining different species at different ratios on the resulting mixtures.

Viscosity and Thermodynamics of Viscous Flow

Table 3 summarizes the experimental values of the kinematic viscosity for the three-component mixtures {ED(1) + ME (2) + DME(3)}. These values complete the relevant data set for the ternary solvent system on the whole, by considering the binary subsystems previously studied [31-33].

Following literature suggestions [34], we choose to measure the kinematic viscosity (ν/cSt) instead of absolute one (η/cP), the latter quantity being related to ν and density (ρ/g cm^{-3}) of the system by the equation

$$\eta = \nu\rho \tag{7}$$

This way, the experimental uncertainties affecting ν values are limited to this fundamental property only, while the implicit uncertainties in η measurements can be due to different and overlapping effects affecting both ν and ρ data contemporarily. In general, η represents an important parameter involved into many processes design calculations, and which is strongly sensitive to specific intermolecular interactions appearing in real systems. Unfortunately, ν and ρ are also and both independently sensitive to these specific interactions, and therefore it should be preferably recommended to researchers to provide these quantities separately instead of η values, to avoid overcoming effects which can lead to mismatching interpretations of the experimental evidences.

Viscosity of real or multicomponent systems, as far as its dependence from temperature and composition, are essential aspects in theoretical and applicative researches where mass transport phenomena play a fundamental role. Unfortunately, the lack of general laws describing such phenomena, in connection with the unexaustiveness of the relevant theories has involved, first, the formulation of empirical and semi-empirical equations (recommended for interpolation purposes) and, secondarily, publication of a huge amount of data collections with the aim to reach the best generalisation [34,35].

Bearing this in mind, the dependence from temperature (T/K) has been studied by applying the Vogel-Tamman-Fulcher equation (VTF) [36]:

$$\ln \nu = \frac{a}{T - T_0} + b \tag{8}$$

for which a quasi-theoretical background based on free volume effects exists, and where a, b and T_0 are the adjustment parameters. VTF model has been found to describe very well the behaviour of many simple liquids (such as mercury, gallium and CCl_4 among others), as far as pure or mixed Newtonian liquids, and true rheometric systems [37]. The goodness of this fit procedure can be evaluated on the basis of the absolute relative deviation ($\Delta\nu$%), calculated as follows:

$$\Delta \nu\% = \frac{100}{n} \sum_{n} \frac{\left| \nu_{exp} - \nu_{calc} \right|}{\nu_{exp}} \qquad (9)$$

where n is the number of experimental points for each liquid mixture.

It should be noted that, in Eq (8), the constant $b = \lim_{T \to \infty} (\ln \nu)$ is a term independent on T, while T_0 is a singular temperature characteristic for each fluid system. A reliable interpretation concerning the T_0 fitting parameter has been proposed in the literature by approaching the limits of the mechanical stability for the supercooled liquids, either Newtonian or rheological fluids, before starting the homogeneous nucleation of the solid phase. In this light, the T_0 term should represent the singular temperature below which the viscosity (or rheometric properties) rapidly increases becoming infinite (for the solid state), and defines the limit field below which the free intermolecular volume is no longer available for the viscous processes. When a rapid increase of viscometric properties occurs in the region of low temperatures, then the onset of a pseudo-glassy state, that preludes the solidification of the system, may be suggested. In this evenness, the condition $\nu \to \infty$ for the selected pure components should be reached at temperatures much higher than those obtained as a fitting parameters (T_o = 142.99 K (ED), 85.01 K (ME), -64.18 K (DME)) [31-33], the melting points (T_{ED} = 261.65 K; T_{ME} = 188.05 K; T_{DME} = 215.15 K) being quite far and very different than the actual T_0 values.

However, in spite of all these considerations, we remember that all the T_0 values obtained for the mixtures pertaining to the three binary sub-systems and the ternary liquids, seem to be characterized by a reasonable internal consistency, since for the examined mixtures they range between the limits represented by the pure species. In the literature, this evident anomalous behaviour for our selected pure components and their mixtures, as far as a variety of other species, has been tentatively explained by invoking molecular association in the liquid state. These phenomenologies also should be responsible for the easy supercooling of the pure components (in particular when hydroxyl substituent groups are present), and coherently with their asymmetric and nonplanar molecular geometry. However, in spite of these remarks about the limits and interpretative difficulties on T_0 parameter, we remember that the availability of Eq (8) for best fitting procedures is warranted when a plot of (ln ν) vs. $(T - T_0)^{-1}$ shows a linear trend over all the temperature range investigated.

Viscosimetric properties of mixed liquids strongly depend also on composition of such real solvent systems. In fact, in the literature it has been largely recognized that viscosity *vs.* composition curves are generally not linear. In particular, these relations show a maximum or a minimum, in the case of not monotonous functions, or none of them, for monotonous functions, or both these features, when "structure making" and "structure breaking" effects in mixed liquids are prevailing or balanced, respectively. Lacking a general law ν = ν(composition), a number of empirical models and theoretical equations have been successfully developed in the past [35] but, unfortunately, their validity, applicability and generalization, are not at all well established in the light of some basilar criteria such as: *i*) ease of employment; *ii*) availability of input parameters; *iii*) range of applicability; *iv*) accuracy in reproducibility of experimental data.

To overcome these problems, we have examined some equations and, between them, we have chosen to correlate each isothermal set (42 values) of the kinematic viscosities with the ternary composition by applying the following polynomial expansion:

$$\ln \nu = \sum_{i=0}^{4} \sum_{j=0}^{4} b_{ij} x_1^i x_2^j \tag{10}$$

This modelling procedure performs very well, giving interpolated values within an average absolute deviation (AAD) $\overline{\Delta\nu} = 0.0068$ cSt (evaluated on the complete data set of 798 experimental points), never exceeding the maximum limit $\Delta\nu = 0.0189$ cSt, and, therefore, we may think that this condition should be valid for any predicted ν value in correspondence of experimental data gaps.

Excess (Deviation) Function

It is since a long time that a passionate discussion involves a great number of researchers worldwide about the physico-chemical description and analytical representation of viscosimetric properties of mixed liquids. This exciting scientific debat starts from the concept that, for all the intensive physico-chemical properties, it is necessary to correctly refer each reasonment to an *ideal* mixed system, this fact implying the fundamental assumption of an appropriated *ideal* mixing law. In fact, it is well understood that any isothermal mixing process in the liquid state, gives a shift of any actual value of a generic property (Y) and the corresponding ideal quantity (Y_{ideal}). Such a shift is called excess function (Y^E) for true thermodynamic properties, or deviation function (ΔY) for extrathermodynamic oncs:

$$Y^E (\text{or } \Delta Y) = Y - Y_{\text{ideal}} \tag{11}$$

Obviously, the term Y_{ideal} is unknown in principle, with the unique consciousness that it is a continuous function of the continuous variable "composition" of mixed systems. As a consequence, we are unable to evaluate and predict neither in sign nor in intensity the Y^E (or ΔY) quantity for any given liquid mixture, and the experimental measurements only permit us to know such properties.

Despite of the giant effort given by the researchers to individuate such law, at present time we don't still have use of it and, therefore, we can only make reference to a choice of a model that, reasonably, can approach such law.

Table 4. Sign and mole fraction (x_2) of maximum deviation of kinematic viscosity ($\Delta\nu$) for the different ideal viscosity models for the three binary sub-systems containing ED, ME and DME at 25 °C

Binary system	$\Delta\nu_a$	$\Delta\nu_g$	$\Delta\nu_h$
ED(1) + ME(2)	< 0 min at $x_2 \cong 0.65$	< 0 min at $x_2 \cong 0.75$	> 0 max at $x_2 \cong 0.85$
ED(1) + DME(2)	< 0 min at $x_2 \cong 0.35$	< 0 min at $x_2 \cong 0.70$	> 0 max at $x_2 \cong 0.95$
ME(1) + DME(2)	< 0 min at $x_2 \cong 0.5$	< 0 min at $x_2 \cong 0.5$	> 0 max at $x_2 \cong 0.25$

In particular, as viscosimetric properties are concerned, it is unclear whether the ideal mixing law should take the form of the weighted arithmetic (Y_a), geometric (Y_g) or harmonic (Y_h) mean of the viscosities of the pure components, each mean properly defined by the corresponding equation:

$$Y_a = \sum_i x_i Y_i \qquad (12)$$

$$\ln Y_g = \sum_i x_i \ln Y_i \qquad (13)$$

$$Y_h^{-1} = \sum_i x_i Y_i^{-1} \qquad (14)$$

At this stage, we are obliged to mention that these relations represent a limited set among many other models that can be safely employed to describe the ideal mixing law [38]. However, looking at this rich literature panorama, one can deduce that the above selected equations should be the most authoritative and widely used for these aims.

Now, with so many different relationships, all being equally practicing, the question is: which is the right choice to describe the *ideal* mixing process? The response is very hard to find *at priori*, but a deeper examination of the results obtained by applying the three different models to the three binary subsystems can simplify the problem and can help us to make the final choice. Table 4 summarizes some elements particularly significant, such as the sign of the excess (deviation) quantity, and the position of the maximum deviation for each binary containing ED, ME and DME.

At first, one may observe that the harmonic function provides always contradictory results. In fact, the relevant $\Delta\nu_h$ data are always > 0, despite the destructuring features both for ME and DME components towards the ED species, and by mixing themselves, as established by studying different thermophysical properties [29,39,40] and applying NMR spectroscopic techniques [23]. Accordingly, the sign of $\Delta\nu$ should be < 0, as given by the arithmetic and geometric ideal mixing rules. Furthermore, the fact that the maximum deviation of different experimental properties in {ED (1) + DME (2)} mixtures generally lie at $x_2 \cong 0.35$ suggests consitency with the behaviour of our $\Delta\nu_a$ data.

All this bearing in mind, we have chosen the arithmetic mean (Y_a) among the other relations, in order to describe the *ideal* behaviour to study the viscosity of this ternary system. Then, each isothermal set of $\Delta\nu$ values (42 data) has been interpolated by applying a relation likes to equation (12), written in the proper form, and the goodness of this fitting procedure is ascertained on the basis of an average uncertainty $\overline{\Delta(\Delta\nu)} = \pm$ 0.0110 cSt, evaluated on the 798 experimental points, and never exceeding the upper limit $\Delta\nu = \pm$ 0.0228 cSt at -10 ^{0}C.

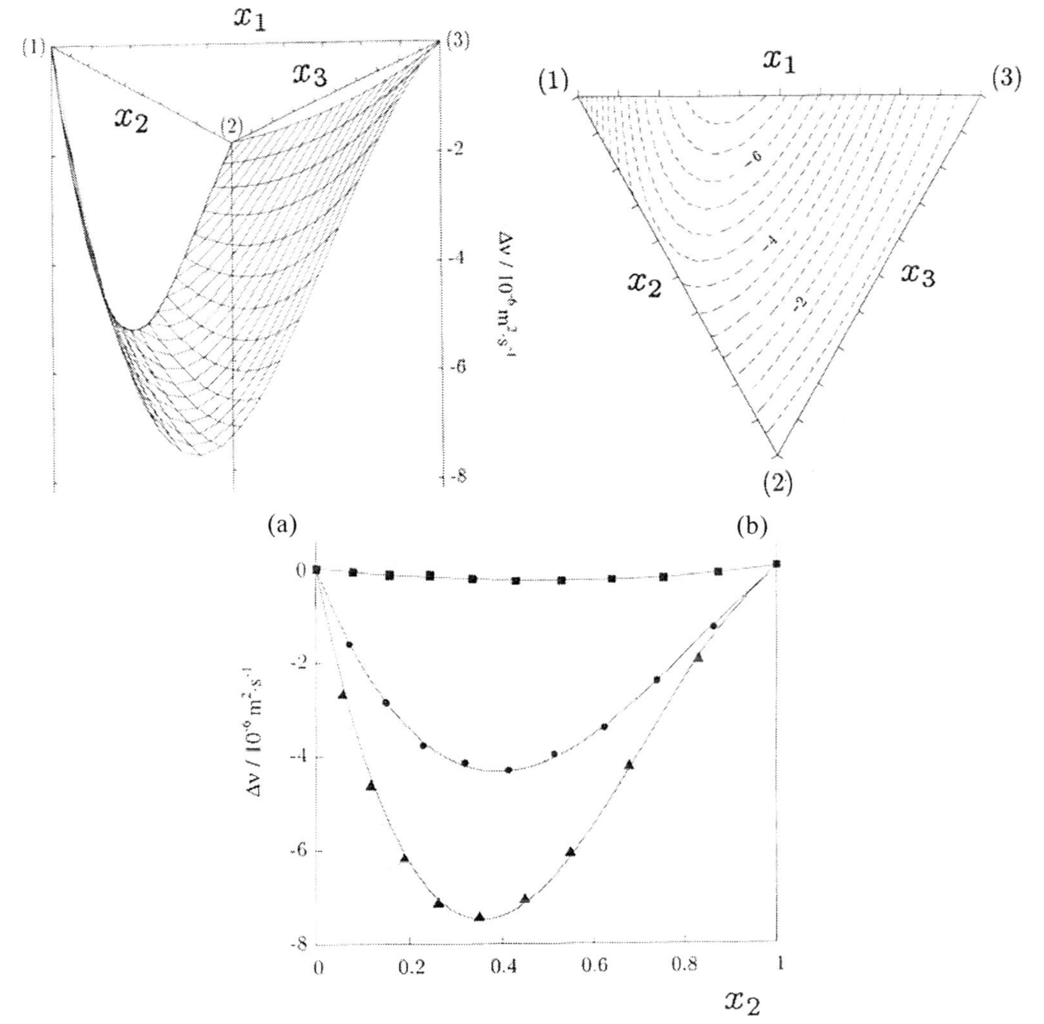

Figure 3. (*a*) Pictorial view of the kinematic viscosity deviation function ($\Delta\nu$) surface at 25 °C for the ternary mixtures {ED(1) + ME(2) + DME(3)}. (*b*) Computer generated contour diagram showing lines of constant $\Delta\nu$ on a liquid – mole fraction (x_i) grid. (*c*) Trends of the $\Delta\nu$ quantity with changing x_i of the three binary subsystems: ● {ED(1) + ME(2)}; ▲ {ED(1) + DME(2)}; ■ {ME(1) + DME(2)}.

As an example of the obtained results, Figure 3 reproduces the trend of the quantity $\Delta\nu = \Delta\nu(x_i)$ at 25 °C. The same trend is almost valid at each temperature in the investigated range $-10 \leq t/^{0}C \leq 80$. This figure has been obtained by plotting the $\Delta\nu$ quantity fitted to the smoothing equation (2) written in the proper form.

A visual inspection of these figures seems to suggest that ternary adducts should be absent in these mixtures. However, these minima are well detectable in the function located along the binary axes. After the very elegant review of Marcus [41], about the solvent structuredness, and on the base of these experimental evidences, we can suggest that we are in presence of a very effective structure-breaking processes with respect to the pure ED component, with a significant structure-making contribution responsible of the appearance of microphases due to solvent-cosolvent adduct and clusters formation phenomena. In addition, we can establish that the most stable solvent – cosolvent adducts formed in this ternary system are in the sequence 2ED·1DME > 2ED·1ME > 1ME·1DME, giving these complex moieties the maximum ability in destructuring the liquid texture.

Thermodynamics of Viscous Flow

Viscosity data can be treated from a thermodynamic point of view according to Eyring's theory [42,43]:

$$\nu = \frac{hL}{\sum_i x_i M_i} \exp(\Delta G^* / RT) \tag{15}$$

where h is the Planck constant, L the Avogadro number, M_i the molar mass of i-th species, R the universal gas constant, T the absolute temperature and ΔG^* is the Gibbs activation energy of the viscous flow.

The Eyring's interpretative model of the viscous flow describes the molecules "jumping" between adjacent laminar layers and moving under a shear stress in a stationary steady state of liquid environment. The field forces stress the molecule to occupy a vacant site of one layer overcoming a free energy barrier.

Combining equation (15) with the fundamental thermodynamic relationship $\Delta G^* = \Delta H^* - T\Delta S^*$, and rearranging, we obtain:

$$R \ln\left(\nu \frac{\sum_i x_i M_i}{hL} \right) = \frac{\Delta H^*}{T} - \Delta S^* \tag{16}$$

where ΔH^* and ΔS^* are enthalpy and entropy of viscous flow, respectively. These quantities can be evaluated from tangent slope and intercept for each curve at each temperature, if the left hand-side term is plotted against $1/T$. As an example, Figure 4 reproduces the trend of Eq. (16) for the three pure components. As depicted in this figure, the trend for pure DME is almost perfectly linear (correlation coefficient $r = 0.9996$), over a relatively wide temperature range. With increasing ED and ME mole fractions in the liquid mixtures, the trend assumes a more pronounced curvature, with the effects of both the hydroxylated cosolvents becoming the prevailing ones. These findings suggest that ΔH^* and ΔS^* are temperature independent in pure DME, while the contrary happens in the case for pure ED and ME species, as well as for all the other mixtures.

A fitting-polynomial procedure can be conveniently applied as an alternative and more suitable approach to the above cited graphical method in evaluating ΔH^* and ΔS^* [44]:

$$R\ln\left(\nu\frac{\sum_i x_i M_i}{hL}\right) = a_0 + \frac{a_1}{T} + \frac{a_2}{T^2} + \frac{a_3}{T^3} = \sum_{j=0}^{3}\frac{a_j}{T^j} \tag{17}$$

where a_j are adjustment coefficients, and

$$\Delta H^* = a_1 + \frac{2a_2}{T} + \frac{3a_3}{T^2} = \sum_0^j j\frac{a_j}{T^{j-1}} \tag{18}$$

$$\Delta S^* = -a_0 + \frac{a_2}{T^2} + \frac{2a_3}{T^3} = \sum_0^j (j-1)\frac{a_j}{T^j} \tag{19}$$

Generally, the third polynomial degree in Eq. (17) is sufficient to warrant a very good approximation and overlapping between experimental and calculated ν data.

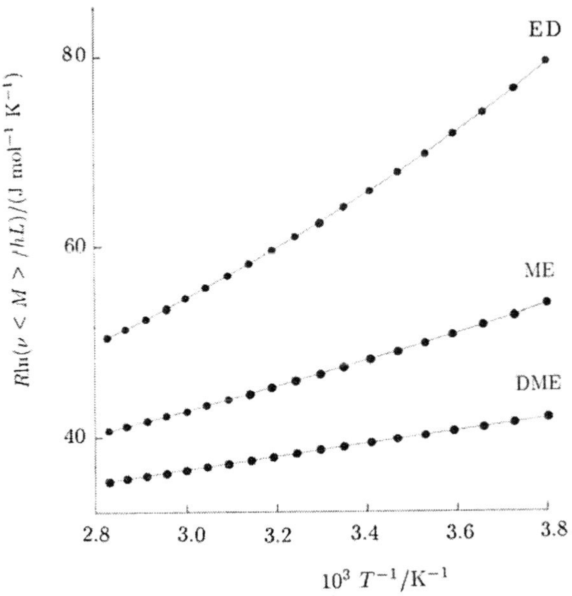

Figure 4. Plots of $R\ln(\nu\langle M\rangle/hL)$ vs. T^{-1}, Eq (16), for the three pure species ED, ME and DME.

As ΔG^* values is concerned, we observe that they are always in the range {11.08 (pure DME at −10 ^{0}C) ≤ ΔG^* / kJ mol^{-1} ≤ 21.09 (pure ED at −10 ^{0}C)}, while the temperature coefficients ($\partial\Delta G^* / \partial T$) are partly > 0 (pure DME, some binaries ED + DME and ME + DME, and some ternaries particularly riched of DME), and partly < 0 (solutions particularly riched of ED).

On the contrary, the temperature coefficients ($\partial\Delta H^* / \partial T$) are always < 0, with the only exception of pure DME. In the case of ΔS^*, the temperature coefficients ($\partial\Delta S^* / \partial T$) are still generally < 0, with the sign inversion (> 0) in pure DME.

The temperature invariance of ΔH^* in pure DME, as well as in other liquids with similar behaviour, as shown by the linear trend in Figure 4, may suggest that the mechanism of viscous flow in this solvent could be a single, thermally activated, process. In other words, DME molecules should scarcely interact with one another, HBD sites being absent, so that specific intermolecular interactions should only consist of much weaker dipole-dipole attractions. These considerations support the hypothesis that, in pure DME, a sufficient number of activated monomers are always under the suitable form to facilitate the viscous flow. On the other hand, in liquid mixtures with hydrogen bonding donating units (such as ED and ME), the flow process could also be ascribed to single activated moieties but, in this case, the flowing entities should be single DME monomers, heteroaggregated adducts between different component molecules to be viewed as single units, and homoaggregated species of the same nature, resembling adducts of homogeneous microphases of the type $(ED)n$, $(ME)m$, etc. All the polymolecular aggregates (both homo- and hetero-aggregates) are characterised by a suitable thermodynamic stability constant, which should be helpful in attributing a microstructure to these mixed liquids; it is reasonable to suppose that these microphases are dependent on time, local environment and temperature. As a consequence, the viscous flow in these mixed systems should be a cooperative intermolecular process, rather than a monoactivated phenomenon involving singular isolated molecules.

As far as ΔH^* is concerned, we notice that this quantity always assumes positive values under any experimental conditions - which is quite reasonable by considering that the viscous flow process requires energy supply - generally increasing with increasing the ED content in the mixture and decreasing at increasing the temperature ($\partial \Delta H^* / \partial T < 0$), the only exception to this trend being pure DME, for which ΔH^* increases with temperature ($\partial \Delta H^* / \partial T > 0$). In conclusion, ΔH^* for this ternary solvent system assumes values in the range 39.5 (pure ED at $-10\ ^0C) \geq \Delta H^* / kJ\ mol^{-1} \geq 6.7$ (pure DME at $-10\ ^0C$).

The ΔS^* values are partly > 0 and partly < 0 (the latter case holds for pure DME and for mixtures that are particularly rich in this component), and always in the interval -16.6 (pure DME at $-10\ ^0C) \leq \Delta S^*/J\ mol^{-1}\ K^{-1} \leq 70.1$ (pure ED at $-10\ ^0C$). The temperature coefficient of this thermodynamic quantity, $\partial \Delta S^* / \partial T$, is always < 0, only except for pure DME. These evidences suggest that the flow mechanism in the {ED (1) + ME (2) + DME (3)} ternary solvent system could be mainly provided for by a transversal motion (orthogonal to the principal direction of the motion) of heteroaggregated adducts, when crossing an energy barrier due to a velocity gradient; a significant contribution of DME monomers has to be taken into account in the corresponding rich-composition region, occurring through an activated state involving primary homo- and hetero-aggregated clusters, or monomeric molecular species [42].

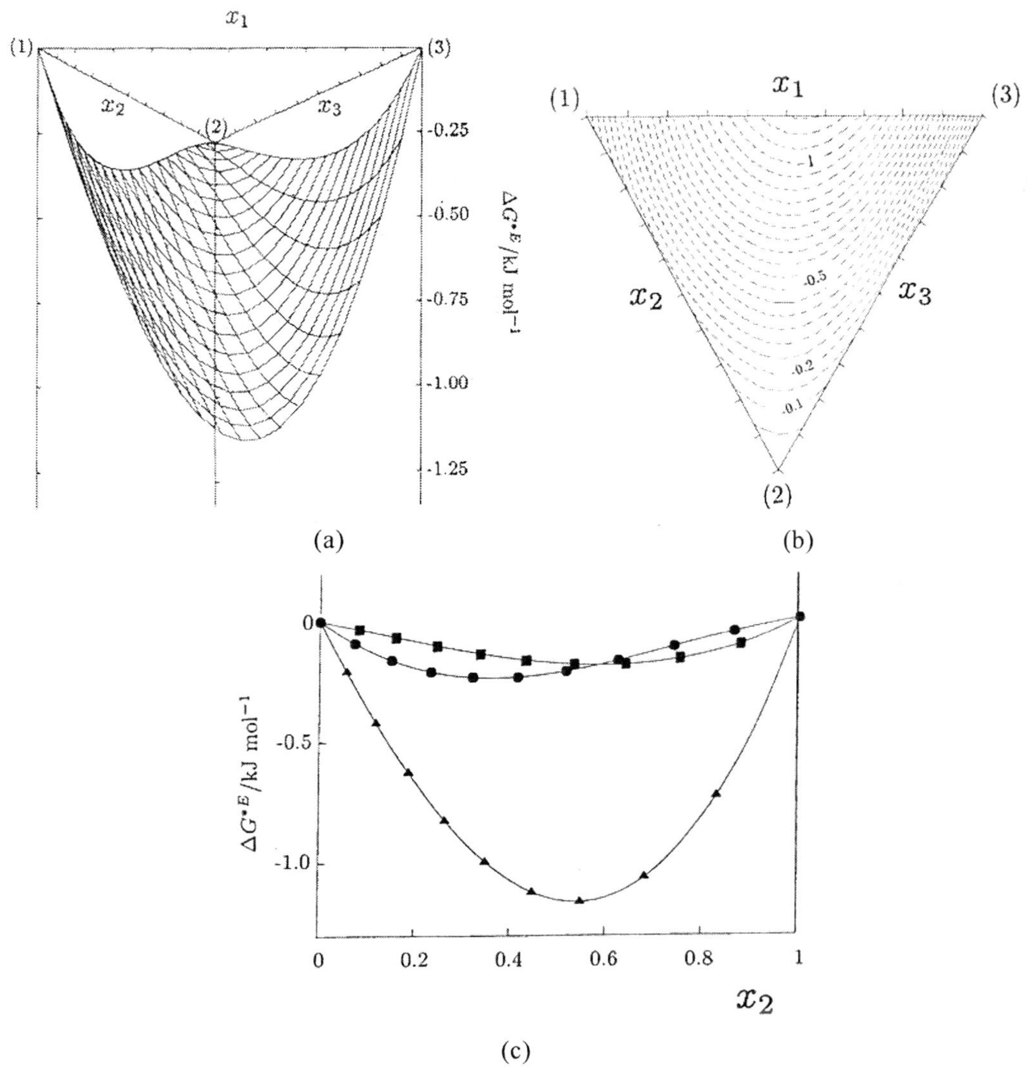

Figure 5. (*a*) Three-dimensional representation of the ΔG^{*E} at 25 °C for the ternary mixtures {ED(1) + ME(2) + DME(3)}. Full-line reticulates represent the surfaces generated by Eq (2). (*b*) Computer generated contour diagram showing lines of constant ΔG^{*E} on a mole fraction (x_i) grid. (*c*) Trends of ΔG^{*E} with changing x_2 of binary systems: ● {ED(1) + ME(2)}; ▲ {ED(1) + DME(2)}; ■ {ME(1) + DME(2)}.

Excess thermodynamic properties of viscous flow (ΔY^{*E}) can be evaluated by applying Eq.ns (11), (12), and then these quantities have been modelled by Eq (2). The results concerning ΔG^{*E} quantities are reproduced in Figure 5.

From a first glimpse of the trends reported in Figure 5 it is possible to note that ΔG^{*E} has always negative values, and this occurs at each experimental condition. It has been outlined in the literature [45,46] that excess molar quantities such as ΔG^{*E} can be taken as a very reliable evidence to check the possible presence of specific intermolecular interactions between different species, as well as to estimate the relevant intensities. A comparison between the three binary subsystems at 25 °C is shown in Figure 5.c; it is clear that the extent of ΔG^{*E}

quantities follows, in general, the sequence {ED + DME} > {ED + ME} ≅ {ME + DME}. This behaviour has been detected at each temperature investigated, even if an increase in the temperature should reflect a consequent decrease in the specific interactions which, in turn, should slacken the aggregated moieties and diminishes the intensity of the deviation function. In agreement with other authors [42,45,46], we definitively estimate that, for $\Delta G^{*E} < 0$, strong short-range intermolecular aggregation phenomena can be operative, due to establish of HBDA specific interections [47], but without networking between different units (or different complex adducts) as observed in ED and ME with *structure-breaking* effects with respect to the pure components. However, in addition to these interactions, the long-range dispersion forces should prevail over any other ones in these liquid mixtures. Finally, on the basis of the evidences of Figure 5, we can conclude that in these ternary mixtures no formation of three-component-adducts occurs, singular points such as minima or maxima being absent in the $\{x_1, x_2, x_3\}$ ternary domain, except for those observed along the binary axes.

Similar informations can be achieved by checking the trends of ΔH^{*E} and ΔS^{*E}.

Relative Permittivity, and Dielectric Properties

The experimental ε values of the three pure species and of the 12 ternary mixtures are listed in Table 5 together with the ternary composition expressed as mole fraction (x_i) of each of the three solvents.

The dependence of the measured ε values on the temperature (T/K) was studied by applying the following relationship [48]:

$$\ln \varepsilon = \sum_{j=0}^{1} \alpha_j T^j \tag{20}$$

where α_j are the empirical adjustment coefficients to be determined by a least – squares method. The above equation, which is available for fitting over a wide temperature range near to the limits $-100 < t\,/\,^0\mathrm{C} < 150$ [49], yields good results, being the average uncertainty associated to the interpolated values equals to $\overline{\delta\varepsilon} = 0.08$, and never exceeding the upper limit ± 0.19 ε units over the whole data set composed by 798 ε values (that is by considering the 3 pure species, 27 binaries [50-52] and 12 ternaries containing ED, ME and DME).

In the literature, dielectric constant data have been also reported as a function of temperature conforming to the equation [49]:

$$\varepsilon = \sum_{j=0}^{1} \alpha'_j T^{-j} \tag{21}$$

where α'_j are the new empirical parameters to be determined. This relation provides a set of calculated ε values within to an AAD, $\overline{\delta\varepsilon} = 0.64$ over all the 798 experimental data of ε. As it appears on the basis of the above-cited AAD values, it is possible to assert that Eq. (20) is more effective than Eq. (21), when applied to our data set. In fact, to obtain the same performance level from the latter, it is necessary to introduce a further term (j = 2) into the polynomial expansion (the third empirical coefficient α'_j of Eq. (21)). For these reasons, and according to many literature suggestions [53], it is confirmed the validity and the better availability of Eq. (20) in ε data reduction processing of {ED(1) + ME(2) + DME(3)} ternary solvent system.

The distributions of the 42 isothermal points of ε in the ternary domain $\{x_1, x_2, x_3\}$ were examined by graphic inspection, in order to check the internal consistency, before compiling the composite sets. Because of the evident non-linearity of such plots, the $\varepsilon = \varepsilon(x_i)$ correlation equations were expected to account for this evidence. In the attempt to find suitable correlation functions for these nonlinear relationships, it is possible to force linear convergence for binary mixtures through a polynomial with high-power terms, such as:

$$\ln \varepsilon = f(x_i) = \sum_{0}^{j} a_j x_i^j \tag{22}$$

Table 5. Ternary composition (x_i) and experimental relative permittivity values (ε) for the {ED(1) + ME(2) + DME(3)} solvent system at various temperatures

x_1	1	0	0	0.3192	0.4727	0.1516	0.1871	0.0770	0.1323	0.3742	0.7191	0.2987	0.8588	0.1237	0.6022
x_2	0	1	0	0.5722	0.5143	0.5299	0.3028	0.6925	0.7890	0.3964	0.1957	0.1963	0.0738	0.0924	0.1539
x_3	0	0	1	0.1086	0.0131	0.3185	0.5101	0.2305	0.0788	0.2294	0.0852	0.5050	0.0675	0.7839	0.2439
t / °C	ε														
-10	49.23	20.03	8.68	24.68	30.94	17.56	15.37	17.72	21.41	23.24	35.64	17.06	40.55	11.56	27.96
-5	47.80	19.55	8.51	24.05	30.12	17.12	15.00	17.28	20.86	22.64	34.69	16.65	39.49	11.31	27.22
0	46.65	19.09	8.35	23.41	29.29	16.71	14.65	16.87	20.33	22.05	33.78	16.27	38.48	11.07	26.51
5	45.41	18.63	8.17	22.81	28.51	16.31	14.31	16.47	19.82	21.47	32.86	15.86	37.47	10.82	25.81
10	44.49	18.19	8.03	22.23	27.75	15.93	13.99	16.07	19.32	20.94	32.03	15.49	36.53	10.60	25.13
15	43.17	17.75	7.85	21.64	26.97	15.57	13.69	15.70	18.82	20.43	31.18	15.13	35.59	10.38	24.49
20	42.14	17.35	7.70	21.09	26.27	15.17	13.35	15.31	18.35	19.89	30.34	14.76	34.65	10.15	23.86
25	40.97	16.94	7.54	20.53	25.55	14.82	13.05	14.95	17.88	19.38	29.55	14.42	33.77	9.93	23.21
30	39.50	16.54	7.40	19.99	24.86	14.46	12.75	14.58	17.42	18.89	28.77	14.08	32.88	9.72	22.60
35	38.84	16.17	7.25	19.47	24.18	14.12	12.45	14.25	16.99	18.39	28.00	13.73	32.04	9.51	22.02
40	37.99	15.76	7.12	18.98	23.53	13.78	12.16	13.90	16.57	17.91	27.24	13.41	31.20	9.30	21.44
45	36.98	15.38	6.98	18.44	22.87	13.44	11.88	13.55	16.13	17.46	26.51	13.09	30.39	9.10	20.88
50	36.07	15.02	6.83	18.00	22.29	13.13	11.61	13.25	15.74	17.00	25.81	12.75	29.58	8.91	20.33
55	35.14	14.68	6.70	17.51	21.65	12.82	11.34	12.94	15.34	16.57	25.13	12.46	28.84	8.72	19.79
60	34.30	14.32	6.57	17.05	21.09	12.50	11.07	12.61	14.94	16.15	24.47	12.16	28.09	8.53	19.28
65	33.44	14.05	6.46	16.58	20.49	12.20	10.81	12.32	14.56	15.70	23.80	11.85	27.36	8.35	18.76
70	32.68	13.71	6.31	16.16	19.93	11.91	10.54	12.04	14.21	15.29	23.17	11.55	26.64	8.15	18.26
75	31.84	13.34	6.21	15.75	19.43	11.62	10.30	11.74	13.85	14.91	22.55	11.27	25.96	7.98	17.79
80	31.00	13.01	6.09	15.30	18.87	11.33	10.07	11.45	13.47	14.53	21.95	11.01	25.28	7.81	17.32

Although linearised polynomials are commonly used to fit experimental data, it has been demonstrated that rational functions hold several advantages [54]. The most general form of the rational function $\varepsilon = f(x_i)$ is represented by the ratio of the two polynomials, expressed by the following relationship:

$$\ln \varepsilon = \frac{a_0 + a_1 x_i + a_2 x_i^2 + ... + a_p x_i^p}{b_0 + b_1 x_i + b_2 x_i^2 + ... + b_q x_i^q} \tag{23}$$

where a_j and b_j are adjustment coefficients. In order to reduce the cumbersomness of this equation, some additional restrictions can be introduced:

- x_i varies within the limits $0 \leq x_i \leq 1$, suggesting that a likely optimum condition corresponds to $p = q$ [54];
- the polynomial in the denominator cannot assume zero value.

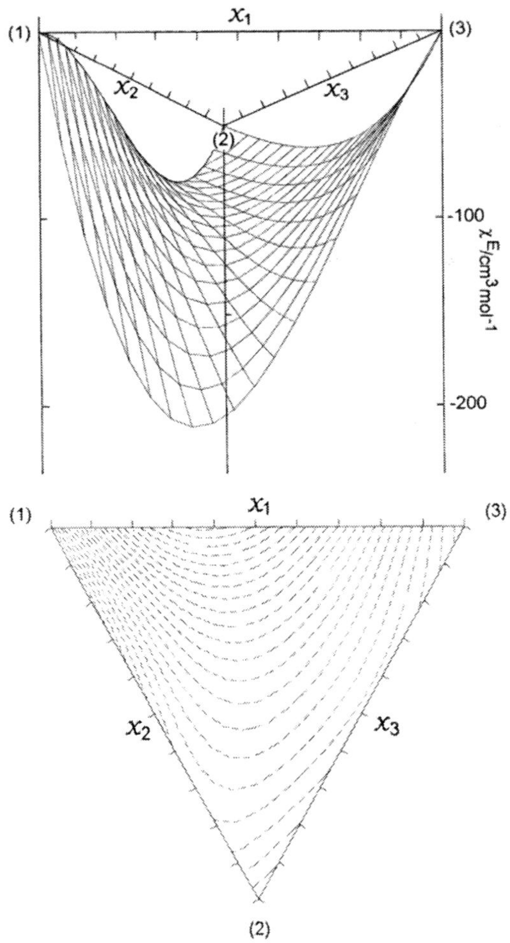

Figure 6. Excess molar dielectric susceptibility (χ^E/cm^{3}mol^{-1}) for the {ED(1) + ME(2) + DME(3)} ternary system at 25 °C, and relevant computer-generated isolines at constant property.

A more restricted form of the rational function (23) is obtained when a_0 is fixed to 1, and only the first-power terms in x_i are considered:

$$\ln \varepsilon = \frac{1 + a_1 x_i}{b_0 + b_1 x_i} \tag{24}$$

Now, it is possible to note that for binary mixtures, equation (24) must reproduce the values of ε_1 and ε_2 at the limits of $x_1 = 1$ and $x_1 = 0$; this condition is actually satisfied when $x_i = x_1$ and $b_0 = (\ln\ \varepsilon_2)^{-1}$, or vice versa. Only the values of the two adjustable parameters a_1 and b_1 must be hence determined. In an effort to extend the validity of this last relation to ternary mixtures too, we applied a modified equation (24) in the form [55]:

$$\ln \varepsilon = \frac{1 + a_1 x_1 + a_2 x_2}{b_0 + b_1 x_1 + b_2 x_2} \tag{25}$$

where $b_0 = \left(\sum_{i=1}^{3} x_i \ln \varepsilon_i \right)^{-1}$.

Similarly to equation (24), equation (25) should also allow the computation of ε_i values for the pure solvents at the respective limits $(x_1, x_2, x_3) = 1$; this way, only 4 adjustable coefficients have to be computed to interpolate each isothermal experimental data set for a given three–component solvent system. Equation (25) has been applied for the calculations at any composition of the ternary solvent system, also taking into account the previously published binary data on {ED(1) + ME(2)} [50], {ED(1) + DME(2)} [51], and {ME(1) + DME(2)} [52] systems; each experimental $\varepsilon(x_i)$ value, for $0 \le x_i \le 1$ $(1 \le i \le 3)$, has been given the same statistical weight in the fitting procedure.

Equation (25) reproduces the experimental data with an average deviation $\overline{\Delta\varepsilon} = 0.19$ over all the 798 experimental values, always assuming values lower than 0.37. This last conclusion confirms the validity of the proposed equation and justifies the choice of only first-power terms for the variable quantities.

In order to accelerate the interpolation procedures, it is possible to combine equations (20) and (25). By algebraic manipulation, the following expression is obtained:

$$\ln \varepsilon = (\mathrm{T}, x_i) = \frac{\sum_{h=0}^{1} \sum_{k=0}^{1} \sum_{l=0}^{1} \gamma_{hkl} \mathrm{T}^h x_1^k x_2^l}{\left(\sum_{i=1}^{3} x_i \ln \varepsilon_i \right)^{-1} + \sum_{i=1}^{2} \delta_i x_i} \tag{26}$$

where γ_{hkl} and δ_i are empirical fitting coefficients. We remind that $(k + l)$ can assume either 0 or 1 value; this is the only condition required to ensure perfect correspondence of the condensed form (26) to the product of single equations accounting for the dependence of ε on T and x_i, respectively.

On the whole, this relationship can be safely employed to obtain interpolated values covering the experimental data gaps, the average uncertainty $\overline{\Delta\varepsilon}$ being equal to 0.39 over the whole investigated temperature ($-10 \leq t\,/\,^0C \leq 80$) and composition ($0 \leq x_i \leq 1$) ranges.

The Deviation (Excess) Function

Trying to define the *ideal* composition dependence of the dielectric properties of mixed associated solvents, starting from the pioneering and fundamental theories of Debye, Onsager, Frohlich and Kirkwood, and the statistical re-elaboration of Winkelmann and Quitzsch [56,57], Davis and Douhéret provided a series of enlightening papers about these arguments [58-60].

Examining the behaviour of a series of different polar and aquomixed binary solvent systems, they argued that the *ideal,* or *regular,* dielectric behaviour closely approximates the condition of mole fraction additivity of the molar dielectric susceptibilities, χ, [defined as follows: $\chi = (\varepsilon-1)\cdot V$] of the pure components:

$$\chi^{id} = (\varepsilon^{id} - 1)V^{id} = \sum_{i=1}^{n} x_i V_i (\varepsilon_i - 1) \tag{27}$$

where the symbols V_i, x_i and ε_i have their usual meaning, and ε^{id} represents the *ideal* dielectric constant of the mixture. The term V^{id} would be assumed as the mole fraction adduct of V_i $\left(V^{\mathrm{id}} = \sum_i \frac{x_i M_i}{\rho_i} \right)$.

As a consequence, we have evaluated the excess counterpart $\Delta\chi$ by applying the eqn.s (11) and (12) written in the proper form. Figure 6 shows the plots of $\Delta\chi$ as a function of the ternary composition domain $\{x_1, x_2, x_3\}$, at 25 ^0C, described in an analytical form by equation (2).

At a molecular level, and as previously outlined, when strong deviations from linearity in Y^E (or ΔY) are present, either a hypothesis of dipolar association between unlike molecules or an increase in dipolar association by one component itself to form self-aggregates of lower dipole moment [61-63], could be suggested in order to represent the experimental behaviour obtained for {ED(1) + ME(2) + DME(3)} ternary mixtures.

A careful examination of Figure 6 shows that no stable three-component adducts are probably formed in these nonelectrolytic solutions, being absent any singular point such as relative minimum or maximum in the ternary domain. However, it is possible to hypothesize that binary adducts may be formed through specific solvent – cosolvent interactions by hydrogen bonding or dipolar interactions of any kind [61-63]. In particular, the most deviating behaviour of $\Delta\chi$ is observed in {ED(1) + DME(2)} binaries, while the maximum deviations in the other binary subsystems {ED(1) + ME(2)} and {ME(1) + DME(2)} are comparable in magnitude at 25 ^0C.

The Molar Polarization

Some structural variations in the liquid mixtures, occurring as a consequence of the specific intermolecular interactions among the different species, can be studied by investigating a dielectric – related property: the molar polarization. It is well known that the total polarization consists of three contributions: (i) one from permanent dipoles related to their respective orientation (for polar species, this part should be the most significant contribution); (ii) one from atomic polarization (or deformation polarization), which should contribute to about 20 – 30% to the total polarization; (iii) the last one from electronic polarization, which is generally assumed to be ≈ 10% of the total.

According to the literature suggestions [58-60], starting from ε values it is possible to evaluate the molar orientational polarizability P_{or}, by applying the following relationship:

$$P_{or} = \frac{(\varepsilon - \varepsilon_\infty)(2\varepsilon + \varepsilon_\infty)V}{\varepsilon(\varepsilon_\infty + 2)^2} \tag{28}$$

where V is the apparent molar volume of the real system $\left(V = \frac{\sum_i x_i M_i}{\rho}\right)$ at each experimental condition, and ε_∞ is the limiting high-frequency permittivity of the medium.

When dealing with polar species, it is a common practice to choose $\varepsilon_\infty \cong 1.1 \cdot n_D^2$ (n_D = refractive index at the Na-D line) at each temperature, this approximation being acceptable [41] for the most common solvents. To reduce the cumbersomness of the formalisms along the text, the subscript P_{or} will be abandoned and the symbol P will be recursively used even retaining the same identical meaning. The molar orientational polarization data of this solvent system vary in the range 556 (ED at 0 °C) $\geq P/\text{cm}^3\ \text{mol}^{-1} \geq$ 123 (DME at 70 ^{0}C).

When referring to a hypothetical ideal mixture of polar – associated liquids, the equation above takes the form:

$$P^{id} = \frac{(\varepsilon^{id} - \varepsilon_\infty)(2\varepsilon^{id} + \varepsilon_\infty)V^{id}}{\varepsilon^{id}(\varepsilon_\infty + 2)^2} \tag{29}$$

where ε^{id} might be derived from Eq. (27), and the high-frequency dielectric constant, ε_∞, for an n – component solvent system would be estimated as follows [59]:

$$\varepsilon_\infty^{id} = \sum_{i=1}^{n} \frac{x_i M_i \varepsilon_{\infty,i}}{\rho_i} \sum_{i=1}^{n} \frac{\rho_i}{x_i M_i} \tag{30}$$

The P data have been used to calculate the excess function P^E, by using Eqn.s (11) and (12) in the proper form. The deviation quantities have been smoothed by using eq (2), and the results at 25 ^{0}C are plotted in Figure 7.

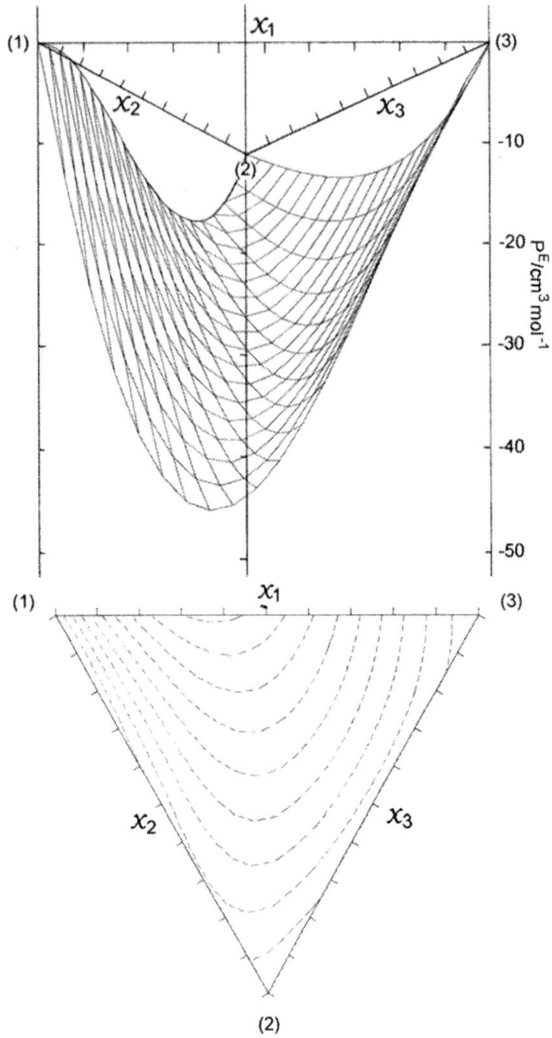

Figure 7. Excess molar (orientational) polarization (P^E/cm^{3}mol^{-1}) for the {ED(1) + ME(2) + DME(3)} ternary system at 25 °C, and relevant computer-generated isolines at constant property.

It is evident that P^E assumes values always negative, under all experimental conditions, and this trend is only slightly sensitive to the temperature. These observations are perfectly coherent with the V^E data of the same mixtures previously described. In other words: the packing degree for these ternaries is increased by the mixing process and, as a consequence of the $P^E < 0$ values, the orientational polarization should decreases when dipolar characteristics of the medium are depressed, that is when strong interactions between component molecules occur.

Molar polarisation (P/cm^3 mol^{-1}) and the related excess quantity (P^E/cm^3 mol^{-1}) represent a molar volume and hence are additive properties. For these properties, it appears to be useful to apply an algebraic manipulation in order to obtain the relevant partial molar quantities $\overline{P}_i$ and $\overline{P}_i^E$. Therefore, to gain further insight into the behaviour of these ternary mixtures, we draw plots of the partial excess molar polarisation of the components, being the $\overline{P}_i^E$ quantities calculated from:

$$\overline{P}_i^E = P^E + (1 - x_i)\frac{\partial P^E}{\partial x_i} \quad (31)$$

Partial molar polarisation $\overline{P}_i$ can be evaluated as follows:

$$\overline{P}_i = P_i + \overline{P}_i^E \quad (32)$$

where P_i is the molar polarisation of i-th pure component at each temperature. As previously outlined for V^E, from the relation (2), and using the proper fitting coefficients at each temperature, the quantity $\partial P^E/\partial x_i$ is obtained and, as a consequence, using equation (32), the $\overline{P}_i$ values can be determined for each experimental condition. As an example, Figure 8 depicts the $\overline{P}_i^E$ quantities in the ternary domain, at 25 ^{0}C. As it can be seen, the values of $\overline{P}_i^E$ are generally negative at the selected temperature for the three components, with only minor exceptions for ME at infinite dilution in ED (Figure 8.b).

It must be remembered that in this type of analysis, the most significant aspects should be related to the trend of the functions, rather than to the local and single value. However, with this consciousness and to make more effective the descriptive comparisons, some discussion about these values will be given, in particular by considering the relevant data for each species at infinite dilution into the other components.

We remember that a negative contribution of $\overline{P}_i^E$ should provide for evidence of consistent interactions arising both from hydrogen bonding or dipole-dipole interactions between different components. This should be the case of ED ($\overline{P}_1^E$) at infinite dilution both in ME ($\overline{P}_1^E \cong$ -58 cm^3·mol^{-1}) and DME ($\overline{P}_1^E \cong$ -57 cm^3·mol^{-1}) (Figure 8.a), confirming that this species should be quite depolarised in the stretched pseudo-linear form (the *anti* conformer), making more easy the hetero-clusterisation processes with the cosolvent molecules. To this respect, it is not surprising that the most negative value of $\overline{P}_1^E$ ($\cong$ -98 cm^3·mol^{-1}) is detected along the binary axis (2)-(3) at $x_3 \cong 0.35$ (2 ME: 1 DME): at this stoichiometry, almost all the hydroxil groups of component mixtures are masked because involved in the formation of hydrogen bonding interactions, and only some residual less polar active sites [-O-H···O(CH_3)-] can be still active to contribute to $\overline{P}_1^E$.

On the other hand, the trend for DME ($\overline{P}_3^E$) at infinite dilution is perfectly complementary towards ED and ME cosolvents (Figure 8.c). In fact, looking at these figures, we can evict that DME undergoes sensible specific hetero-cooperative interactions, giving a very high correlative order degree between different molecules, and depressing its orientational polarization ($\overline{P}_3^E \cong$ -28 cm^3·mol^{-1} and –286 cm^3·mol^{-1} at infinite dilution in ME and ED, respectively).

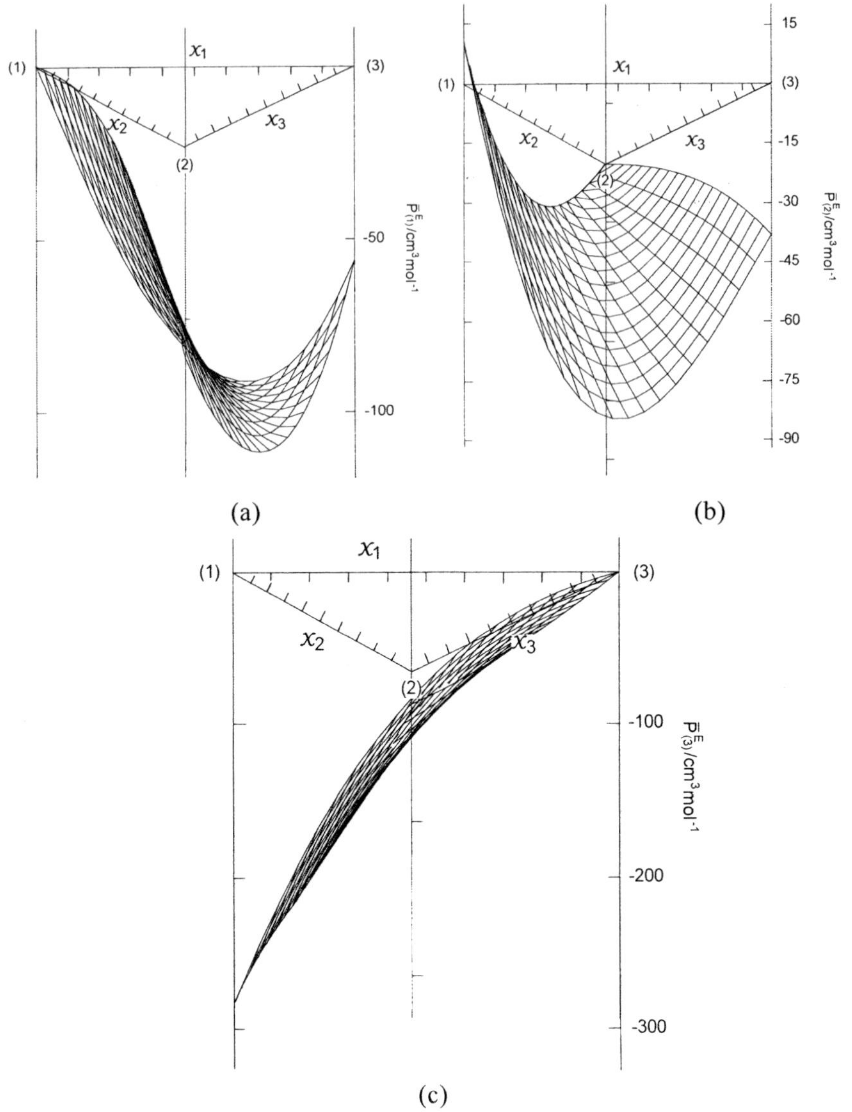

Figure 8. View of the partial excess molar (orientational) polarization ($\overline{P}_i^E$ /cm^3 mol^{-1}) for the {ED(1) + ME(2) + DME(3)} ternary system at 25 °C: (*a*) ED(1); (*b*) ME(2); (*c*) DME(3).

The situation seems to be partially inverted when considering the ME species ($\overline{P}_2^E \cong 11$ cm^3·mol^{-1} and -38 cm^3·mol^{-1} at infinite dilution in ED and DME, respectively). These values seem indicate that at infinite dilution in DME, ME molecules should be strongly engaged to form heteroaggregated adducts *via* hydrogen bonding interactions with the cosolvent.

On the contrary, the ME *gauche* conformer should prevail at infinite dilution in ED (Figure 8.b) confirming that the modest enhancement of the partial molar orientational polarization ($\overline{P}_2^E \cong 11$ cm^3·mol^{-1}) should probably due to the weakness of specific intermolecular interaction with ED molecules because of the lacking elongation of ME [23], and reinforcing to some extent the parallel realignment of the quadrupolar ME moments.

However, as an indication which probably should be valid under all other experimental conditions, the most marked and distorsive effects ($\overline{P}_2^E \cong -85$ cm^3·mol^{-1}) are detected when mixing ME at infinite dilution in equimolar ED + DME binary [$x_1 \cong 0.5$ (1 ED: 1 DME)], where the maximum destructuration of ME component should be reached.

Dipolar Interaction Free Energy

From statistical considerations, Winkelmann and Quitzsch have concluded that the molar dipolar free energy (F_μ) of a pure associated liquid is given by the equation

$$F_\mu = -\frac{RTLg\mu^2(\varepsilon-1)(\varepsilon_\infty+2)}{9k_B T\varepsilon_0 V(2\varepsilon+\varepsilon_\infty)} \tag{33}$$

where g is the Kirkwood correlation factor, and the other symbols have their usual meaning [56,57]. Following that logic, Davis and Douhéret [59] developed a very useful relationship which was adopted for the case of binary mixtures of Onsager liquids:

$$F_\mu = -RT(\varepsilon-1)\sum_{i=1}^{2}\frac{x_i P_i(\varepsilon_{\infty,i}+2)}{V_i(2\varepsilon+\varepsilon_{\infty,i})} \tag{34}$$

The formalism of Eq. (34) is particularly suitable for the extension of its applicability to a generic multicomponent system and we tried to evaluate the quantity $F\mu$ for the ternary solvent system under study. By applying these relations, some interesting results have been obtained.

At first, we observe that F_μ is always negative (attractive interactions are prevailing), with a decreasing trend by increasing temperature: the greatest values have been obtained in pure ED (-22.05 kJ mol^{-1} at 25 ^{0}C), DME shows the most feeble ones (-2.88 kJ mol^{-1} at 25 °C), and ME values are located at the middle between these extremes (-8.30 kJ mol^{-1} at 25 ^{0}C).

The property F_μ is a true additive thermodynamic quantity, and the equation

$$F_\mu^{id} = \sum_i x_i F_{\mu,i} \tag{35}$$

represents an alternative definition of an ideal mixing process, where the molar dipolar free energy of pure i-th component, $F_{\mu,i}$, is evaluated by Eq (35). On the basis of these considerations, it is possible to obtain an excess mixing property:

$$F_\mu^E = F_\mu - F_\mu^{id} \tag{36}$$

which has been evaluated at each experimental condition. The results at 25 ^{0}C are represented in graphical form in Figure 9, where the excess quantity has been fitted to the Eq (2).

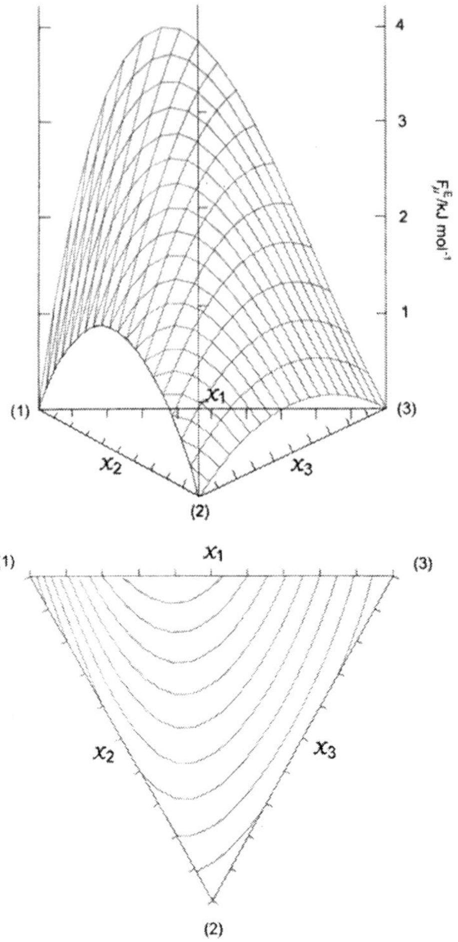

Figure 9. Excess molar dipolar interaction free energy (F_μ^E /kJ mol^{-1}) for the {ED(1) + ME(2) + DME(3)} ternary system at 25 °C, and relevant computer-generated isolines at constant property.

The trend $F_\mu^E = F_\mu^E(x_1,x_2,x_3)$ is always positive, with consistent and greatest deviations along the binary axis in the sequence {ED(1) + DME(3)} > {ED(1) + ME(2)} > {ME(2) + DME(3)}, and reaching the maximum values ~ 4.0 kJ·mol^{-1} at $x_1 \cong 0.65$ (2 ED: 1 DME), ~ 1.3 kJ·mol^{-1} at $x_2 \cong 0.5$ (1 ED: 1 ME), and ~ 0.5 kJ·mol^{-1} at $x_3 \cong 0.5$ (1 ME: 1 DME), respectively. Similar behaviour has been obtained at all other experimental conditions. These findings should indicate that the most stable hetero-adducts would be formed into ME+DME binary subsystem, where less repulsive phenomena should take place between components. On the contrary, greater and positve is F_μ^E, more strong and significant should be the electrostriction phenomena detectable at molecular level. This last observation seems to be perfectly coherent with the relevant results about V^E investigations: the most impressive volumic contraction by mixing have been obtained for ED + DME binaries, suggesting that a more compact liquid structure is perfectly compatible with the appearance of electrostriction (repulsive) phenomena.

To gain further information about this property, we tried to explore the behaviour of the partial molar dipolar interaction energies ($\overline{F}^{E}_{\mu,i}$) for the three components in the mixtures, by applying the following equation

$$\overline{F}^{E}_{\mu,i} = F^{E}_{\mu} + (1 - x_i)\frac{\partial F^{E}_{\mu}}{\partial x_i} \tag{37}$$

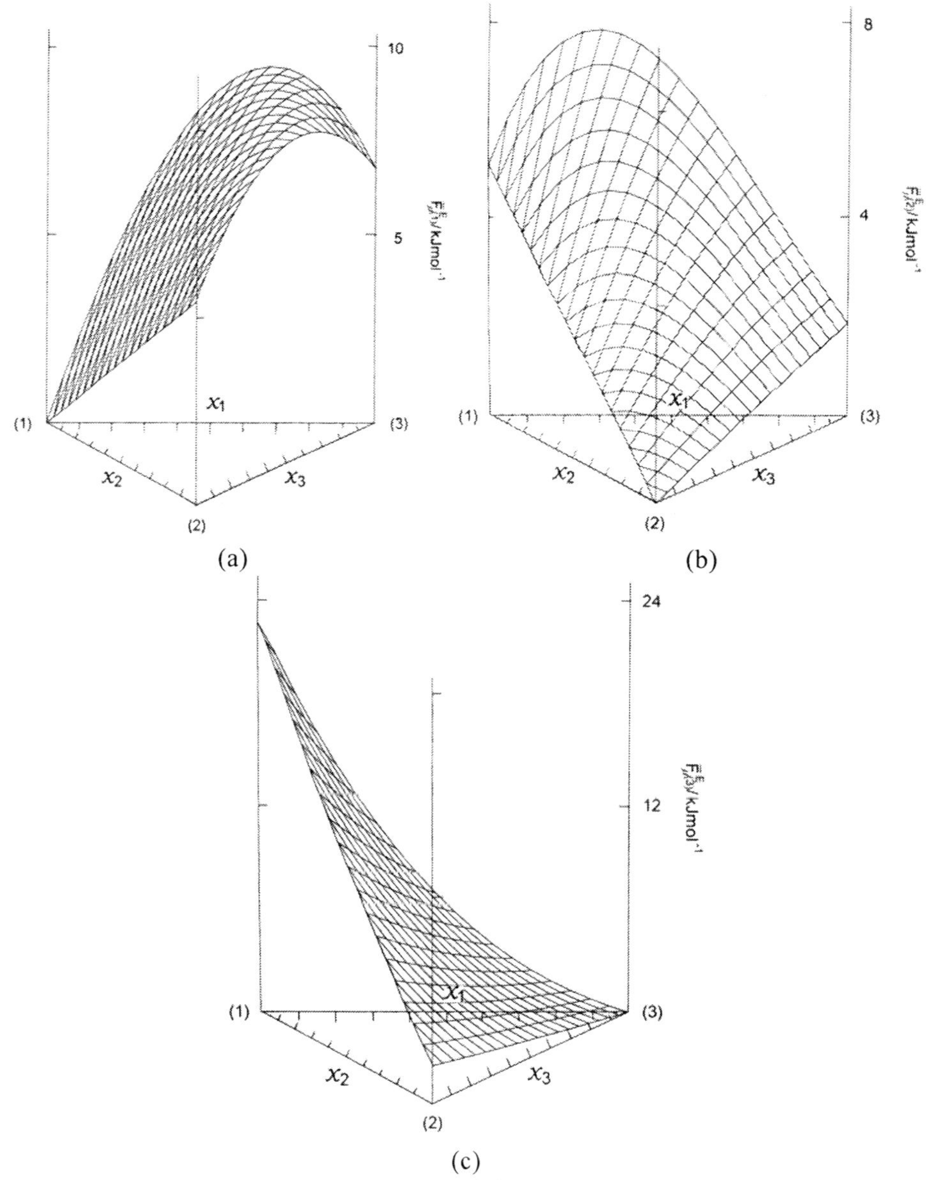

Figure 10. Partial excess molar dipolar interaction free energy ($\overline{F}^{E}_{\mu,i}$ /kJ mol^{-1}) for the {ED(1) + ME(2) + DME(3)} ternary system at 25 °C: (*a*) ED(1); (*b*) ME(2); (*c*) DME(3).

The results are presented in Figure 10. As evident from the figure, $\overline{F}^{E}_{\mu,i}$ is always positive for the three species, at each experimental condition. Therefore, we can generalise that all the components sensibly increase the dipolar free energy by mixing one each other, but the DME molecules undergo the most significant electrostatic stress and energetic perturbation when dispersed at infinite dilution in ED ($\overline{F}^{E}_{\mu,i}$ = 22.6 kJ·mol^{-1} at 25 °C). As a counterpart, ED undergoes the maximum electrostatic perturbation induced by the presence of DME at $x_1 \cong 0.35$ being $\overline{F}^{E}_{\mu,1}$= 9.6 kJ·mol^{-1}, when the mixing ratio becomes 1 ED: 2 DME ($\overline{F}^{E}_{\mu,1}$= 6.8 kJ·mol^{-1} at infinite dilution in DME). As far as ME concerns, we observe that this species is polarly less stressed at infinite dilution in DME ($\overline{F}^{E}_{\mu,2}$= 1.8 kJ·mol^{-1}) rather than in ED ($\overline{F}^{E}_{\mu,2}$= 5.0 kJ·mol^{-1}), differing these values of about 56 %. In addition, the maximum perturbation ($\overline{F}^{E}_{\mu,2}$= 7.9 kJ·mol^{-1}) is detected along the binary axis ED(1) + DME(3) at $x_1 \cong$ 0.65 (2 ED: 1 DME), when electrostriction phenomena involving ED and DME molecules are at the top (see Figure 9).

Refractive Index, and Related Properties

In this section we report some investigations about the values (Table 6) of the refractive index (n) for the ternary mixtures {ED (1) + ME (2) + DME (3)} measured at various temperatures in the $0 \leq t$ / °C ≤ 70 interval (compatibly with the instrumental characteristics and the performance of the refractometer). Some literature relationships, usually applied for binary mixtures and here translated to ternary systems too, have been used to analyze and interpret the experimental data and the derived quantities.

In general, the dependence of n on temperature (T / K) is given by the polynomial relationship:

$$\ln n = \sum_{0}^{k} a_k T^k \tag{38}$$

where a_k are empirical coefficients. It should be remarked that in this polynomial equation only few terms (k = 0,1) are needed to describe the refractive index of several (polar, apolar, protic, aprotic, pure or mixed) liquid substances, even in presence of associative phenomena to whatever extent [64]. The availability of equation (38) for our ternary mixtures is warranted by a regular monotonously decreasing trend of the curves $n = n(T)$ for all the solvents here investigated. Furthermore, the effectiveness of this relation is stated on the basis of an average deviation $\overline{\Delta n}\%$ = 0.006, never exceeding the upper limit of 0.012 for the N = 225 values of Table 6.

Table 6. Values of the molar dipolar free energy (F_μ / kJ mol^{-1}) for the 3 pure components at various temperatures

t / °C	ED	ME	DME
-10	-23.43	-8.85	-3.04
-5	-23.25	-8.77	-3.03
0	-23.06	-8.69	-3.01
5	-22.87	-8.61	-2.99
10	-22.67	-8.54	-2.96
15	-22.47	-8.46	-2.94
20	-22.27	-8.38	-2.91
25	-22.05	-8.30	-2.88
30	-21.83	-8.21	-2.85
35	-21.61	-8.12	-2.82
40	-21.38	-8.03	-2.78
45	-21.15	-7.94	-2.75
50	-20.92	-7.85	-2.72
55	-20.68	-7.75	-2.68
60	-20.44	-7.66	-2.65
65	-20.19	-7.56	-2.61
70	-19.94	-7.46	-2.57
75	-19.68	-7.36	-2.53
80	-19.42	-7.25	-2.48

'Mixing Rules' Equations

In the literature it has been discussed that the refractive index data are strongly affected by thermal effects which, in turn, are reflected in volume effects too [64,65]. As mentioned in the previous sections of this work, the volume properties of ternary liquid mixtures reflect to some extent the composition effects of the systems. Therefore, it is convenient to study the dependence $n = n$(composition) through the true volume fraction ϕ_i, defined by the relation

$$\phi_i = \frac{m_i V_i}{\sum_i m_i V_i} \tag{39}$$

where m_i is the number of moles of i-th species in each experimental condition.

With the aim to obtain further information about predictive ability of some literature relationships when applied to binary solvent systems, tentatively we have checked their adequacy when transferred to these ternary liquid mixtures. These equations, proposed time to time by different authors, are commonly termed "Mixing Rules" (MR) and relate the isothermal refractive properties of the mixed liquids with those of the pure components at each experimental condition.

The selected equations are:

Arago-Biot's equation [66]

$$n = \varphi_1 n_1 + \varphi_2 n_2 + \varphi_3 n_3 \tag{AB}$$

Gladstone –Dale's equation [67]

$$n - 1 = \varphi_1(n_1 - 1) + \varphi_2(n_2 - 1) + \varphi_3(n_3 - 1) \tag{GD}$$

Lorenz-Lorentz's equation [64]

$$\frac{n^2-1}{n^2+2} = \varphi_1\left[\frac{n_1^2-1}{n_1^2+2}\right] + \varphi_2\left[\frac{n_2^2-1}{n_2^2+2}\right] + \varphi_3\left[\frac{n_3^2-1}{n_3^2+2}\right] \tag{LL}$$

Newton's equation [68]

$$n^2 - 1 = \varphi_1(n_1^2 - 1) + \varphi_2(n_2^2 - 1) + \varphi_3(n_3^2 - 1) \tag{NW}$$

Oster's equation [69]

$$\frac{(n^2-1)(2n^2+1)}{n^2} = \sum_{i=1}^{3}\varphi_i\left[\frac{(n_i^2-1)(2n_i^2+1)}{n_i^2}\right] \tag{OS}$$

Finally, the Eykman's rule which is one of the most popular alternative interpolating relationship, taking the form [70]

$$\frac{n^2-1}{n^2+0.4} = \sum_{i=1}^{3}\left[\varphi_i\left(\frac{n_i^2-1}{n_i^2+0.4}\right)\right] \tag{EYK}$$

According to literature suggestions of Hawrylak et al. [71], in these equations, φ_i is a volume fraction expressed on a partial molar volume basis at infinite dilution of component *i*:

$$\varphi_i = \frac{x_i\overline{V}_i}{\sum_i x_i\overline{V}_i} \tag{40}$$

It should be noted that the use of true volume fractions ϕ_i implies to assume not only specific refraction additivity [$r = f(n)/\rho$] when the composition is expressed in weight fractions [72],

Table 7. Ternary composition (x_i) and experimental refractive index values (n_D) for the { ED(1) + ME(2) + DME(3)} solvent system

x_1	1	0	0	0.3192	0.4727	0.1516	0.1871	0.0770	0.1323	0.3742	0.7191	0.2987	0.8588	0.1237	0.6022
x_2	0	1	0	0.5722	0.5143	0.5299	0.3028	0.6925	0.7890	0.3964	0.1957	0.1963	0.0738	0.0924	0.1539
x_3	0	0	1	0.1086	0.0131	0.3185	0.5101	0.2305	0.0788	0.2294	0.0852	0.5050	0.0675	0.7839	0.2439
t / °C															
0	1.43742	1.40992	1.38938	1.41432	1.42141	1.40496	1.40095	1.40593	1.41134	1.41160	1.42482	1.40283	1.42956	1.39452	1.41589
5	1.43601	1.40793	1.38688	1.41252	1.41976	1.40291	1.39879	1.40392	1.40948	1.40972	1.42321	1.40072	1.42802	1.39218	1.41410
10	1.43460	1.40602	1.38439	1.41054	1.41788	1.40074	1.39653	1.40178	1.40739	1.40760	1.42145	1.39845	1.42635	1.38978	1.41214
15	1.43321	1.40404	1.38197	1.40863	1.41611	1.39872	1.39442	1.39978	1.40551	1.40574	1.41975	1.39639	1.42482	1.38751	1.41028
20	1.43185	1.40199	1.37945	1.40672	1.41444	1.39652	1.39212	1.39762	1.40353	1.40372	1.41809	1.39414	1.42326	1.38508	1.40835
25	1.43044	1.40008	1.37702	1.40493	1.41267	1.39452	1.39001	1.39565	1.40168	1.40186	1.41649	1.39207	1.42174	1.38279	1.40656
30	1.42904	1.39810	1.37456	1.40297	1.41094	1.39240	1.38781	1.39354	1.39966	1.39986	1.41477	1.38991	1.41016	1.38045	1.40467
35	1.42761	1.39608	1.37207	1.40103	1.40913	1.39029	1.38561	1.39145	1.39767	1.39788	1.41305	1.38774	1.41854	1.37809	1.40277
40	1.42623	1.39417	1.36963	1.39918	1.40746	1.38821	1.38343	1.38940	1.39576	1.39596	1.41143	1.38561	1.41702	1.37577	1.40093
45	1.42481	1.39220	1.36718	1.39734	1.40571	1.38612	1.38124	1.38734	1.39386	1.39402	1.40978	1.38345	1.41545	1.37343	1.39907
50	1.42341	1.39025	1.36472	1.39542	1.40401	1.38403	1.37906	1.38528	1.39188	1.39205	1.40808	1.38131	1.41385	1.37110	1.39718
55	1.42204	1.38829	1.36229	1.39354	1.40223	1.38199	1.37693	1.38326	1.38995	1.39013	1.40640	1.37922	1.41229	1.36881	1.39535
60	1.42062	1.38626	1.35980	1.39160	1.40041	1.37989	1.37473	1.38119	1.38797	1.38815	1.40469	1.37704	1.41072	1.36645	1.39345
65	1.41917	1.38433	1.35736	1.38966	1.39862	1.37777	1.37254	1.37908	1.38596	1.38616	1.40296	1.37489	1.40907	1.36413	1.39156
70	1.41778	1.38237	1.35493	1.38779	1.39689	1.37570	1.37037	1.37703	1.38403	1.38422	1.40130	1.37276	1.40751	1.36182	1.38970

$$\frac{1}{\rho} f(n) = \frac{w_1}{\rho_1} f(n_1) + \frac{w_2}{\rho_2} f(n_2) \tag{41}$$

but also volume additivity, and then predictions for n are made by using only pure components data.

However, for the sake of completeness, it must be remembered that some other relations, pertaining to two groups, are available for these purposes. The first group contains some equations (Wiener [73], Heller [74], Eyring [29], among other), which are valid for binary systems and not easily transferable to ternary (or upper) mixtures too. The second group contains literature relations (Laplace, Rosen, and other eqs.), some of which have been critically reviewed, among other authors, by Parfitt and Wood [75], Munk et al. [76,77], Segré et al. [78], Brocos et al. [65]. Generally, the applicability of these last models needs the *a priori* knowledge of one (or more) empirical parameters (other than density) for each mixed liquid and, therefore, theirs availability is stronghly limited. It should be noticed that this fact is completely disregarded in all the previously examined 'Mixing Rules' set.

The most conclusive validity test about the relative merits of these models consists in comparing the experimentally measured refractive indeces of the mixtures with those calculated from the refractive properties of the pure species [17,79,80]. All the checked "Mixing Rules (MR)" provide analogous results in each experimental condition, and therefore they are perfectly equivalent for predictive purposes when applied to the {ED(1) + ME(2) + DME(3)} ternary mixtures, though their effectiveness is fairly good as the temperature is lowered.

As a final remark about the selected MR equations, it should be noted that these literature relationships do not contain any fitting parameter and, as a consequence, they are immediately applicable for predictive calculations if the values of the pure species and the mixture compositions are known.

The Deviation (Excess) Function

On the basis of the non-linearity of the relation $n = n(\phi_i)$ (or, alternatively, $n = n(\varphi_i)$), and following literature suggestions [80], it is possible to evaluate the deviation function of the refractive index, Δn, by applying the following equation:

$$\Delta n = n - (\phi_1 n_1 + \phi_2 n_2) \tag{42}$$

where the term $(\sum_i \phi_i n_i)$ should represent the ideal behaviour of the refractive index for a multicomponent solvent system. In this case, the excess property is always < 0, and the results at 25 °C have been plotted in Figure 11. The trend of these curves shows an absolute minimum along the binary axes ED(1)-DME(3) that becomes much deeper as temperature increases, always centred around the abscissae $\phi_1 \cong 0.65$ at each experimental condition. Deviations from ideality, either positive or negative, are indicative of the existence of specific interactions of any kind between components [80], and in many cases provide evidence of solvent-solvent or solvent-cosolvent adducts formation in solution.

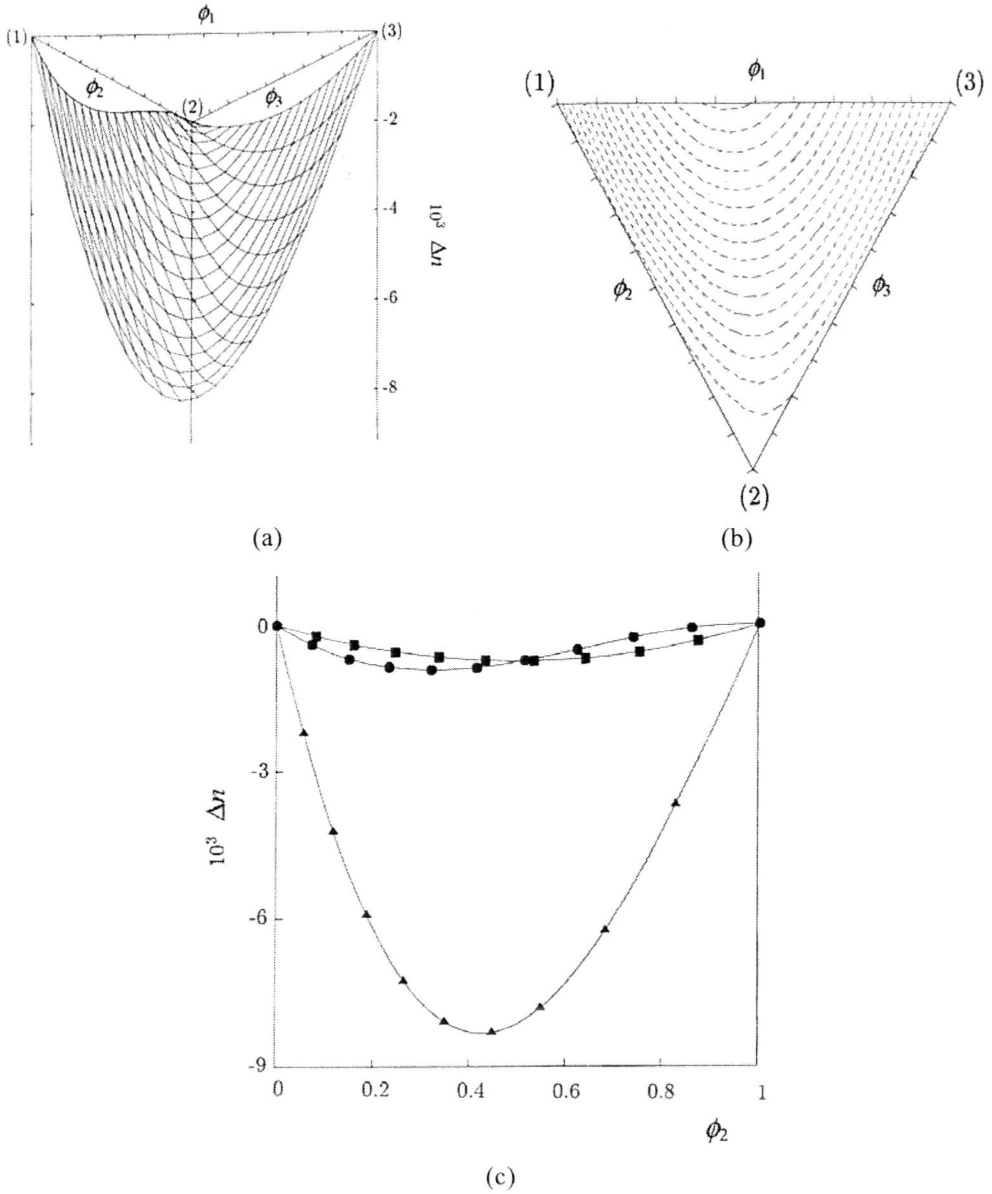

Figure 11. (*a*) Three-dimensional representation of the experimental data of Δn at 25 °C for the ternary mixtures {ED(1) + ME(2) + DME(3)}. Full-line reticulates represent the surfaces generated by Eq (43). (*b*) Computer generated contour diagram showing lines of constant Δn on a liquid – volume fraction (ϕ_i) grid. (*c*) Profiles of the Δn with changing ϕ_i of binary subsystems: ● {ED(1) + ME(2)}; ▲ {ED(1) + DME(2)}; ■ {ME(1) + DME(2)}.

Refractive indices are indirectly related to dispersion forces between molecules, which are due to instantaneous dipole formation as a result of oscillations of electron clouds around the cores of the molecules. As a consequence, the changes in refractive indices are related to the dielectric behaviour of mixtures, also reflecting molecular features (as symmetry, presence of active groups, ecc.) in sign and magnitude. The changes in refractive index of these mixtures containing ED, ME and DME are rather small with composition for the ternary system, and its extreme amounts to 0.53 % of the molar refractive index of the same three-components mixture.

The Molar Refraction

To complete this study, we have considered a further aspect by taking into account a related quantity, i.e. the molar refraction (R / cm^3·mol^{-1}) defined by the equation [64]:

$$R = \frac{n^2 - 1}{n^2 + 2} \cdot \frac{\sum_i x_i M_i}{\rho} \tag{43}$$

The analytical structure of equation (43) may be conveniently represented as a product of two independent functions, taking the general form:

$$R = f(n) \cdot f(V) \tag{44}$$

where $f(n)$ and $f(V)$ are defined as "refractive index function" and "volume function", respectively. Therefore, two well-defined and different contributions concur to modulate the R quantity in each experimental condition [80].

The obtained R data for the {ED (1) + ME (2) + DME (3)} solutions are in the range 14.44 (pure ED at 0 °C) ≤ R/cm^3·mol^{-1} ≤ 24.22 (pure DME at 70 °C). Looking at R data generated by eq. (43), wich in turn are related to the electronic polarization of the mixtures, as an evidence one can obtain a monotonous increase with both temperature and DME mole fraction in ternary composition. Furthermore, the trend is only slightly influenced by temperature; the molar refraction of ED at 0 °C is about 24.8 % lower than that of ME and 39.9 % of DME, and the difference slightly increases with temperature up to a value of 25.0 % and 40.2 % for ME and DME, respectively, at 70 °C. All these observations permit us to state that the higher R values for DME perfectly agree with the highest molecular mass of this component, while the presence of very polar hydroxyl groups into ED and ME skeletal structures, don't compensate the above reported molecular features with respect to cosolvent DME. This way, we can assert that the permanent electrical dipole moments don't significantly contribute to the total polarization of the studied system when irradiated with frequencies of the visible region.

Starting from the R values, the relevant excess quantity (R^E/cm^3·mol^{-1}) has been calculated by applying the equation [65,80]:

$$R^E = R - \sum_{i=1}^{3} x_i R_i \tag{45}$$

and then smoothed to the equation (2) written in the proper form for this excess quantity. The results obtained at 25 °C are shown in Figure 12.

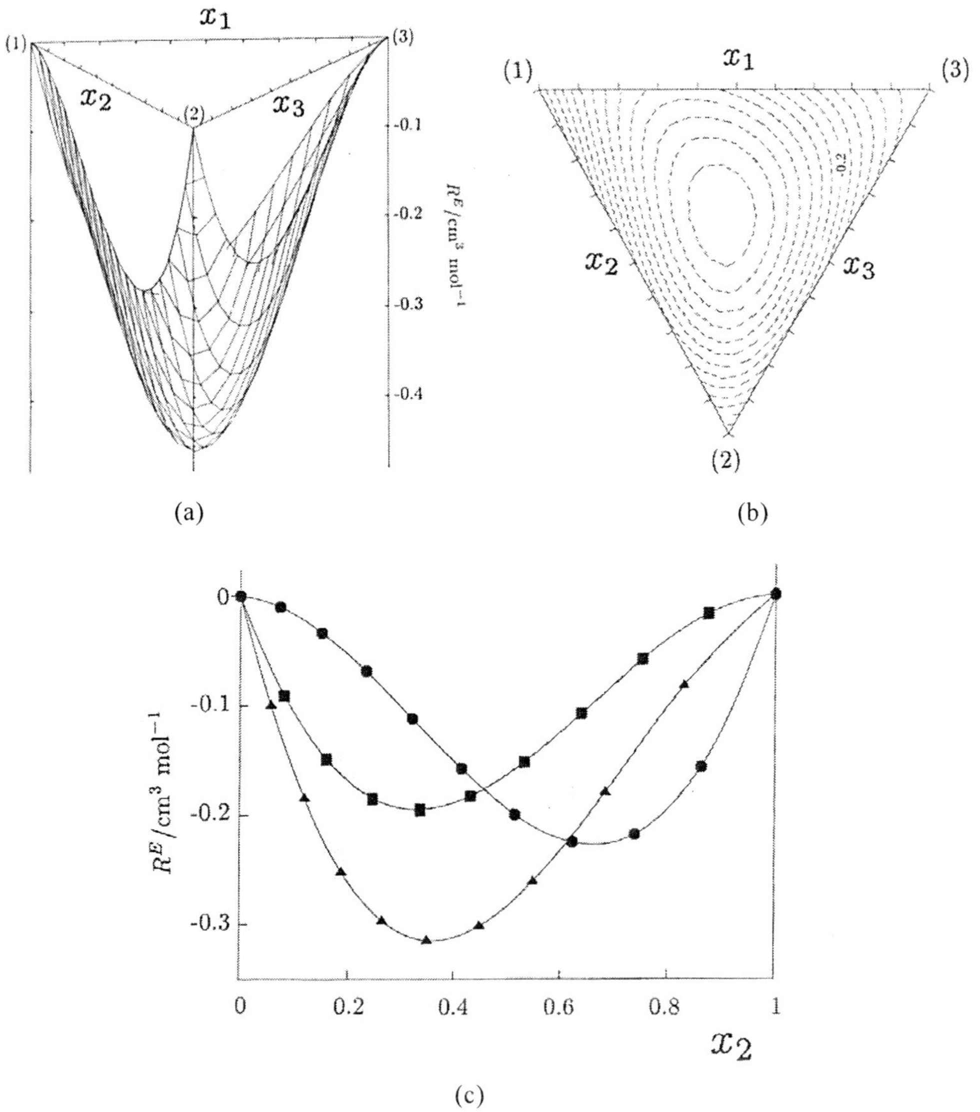

Figure 12. (*a*) Three-dimensional representation of the R^E data at 25 °C for the ternary solvent system {ED(1) + ME(2) + DME(3)}. Full-line reticulates represent the surfaces generated by Eq (2). (*b*) Lines of constant R^E on a liquid – mole fraction (x_i) grid. (*c*) Profiles of the R^E with changing x_i of the three binary subsystems: ● {ED(1) + ME(2)}; ▲ {ED(1) + DME(2)}; ■ {ME(1) + DME(2)}.

The deviations from ideality are always negative, and all the curves (surfaces) show singular points (minima) always positioned at the center of the ternary domain ($x_1 \cong x_2 \cong x_3 \cong 0.33$ (1 ED: 1 ME: 1 DME)), while additional minima are present along the binary axes, centred at $x_1 \cong 0.65$ (2 ED: 1 DME), $x_2 \cong 0.65$ (1 ED: 2 ME), and $x_3 \cong 0.35$ (2 ME: 1 DME) at each temperature, respectively. Furthermore, we observe that an increasing temperature provides a sensible magnification of these singular points.

From a phenomenological point of view, and as previously mentioned, the predominant effect of mixing the selected components, appears to be a consistent lowering of the total polarity of the mixtures and a consequent increasing of the destructuring character of the liquid system. This observation accounts for the behavior very well, and more likely, a strong depolarisation occurs due to hydrogen bonding or other specific intermolecular interactions, and the probable separation of heterogenous microphases is provided.

CONCLUSIONS

A lot of works reported in literature in this field employed binary mixtures as liquid systems. On the other hand, ternary (or upper) solvent mixtures have been only slightly reviewed in spite of their most representative behaviour of real using conditions systems. Ternary systems are very attractive from this point of view, and this great interest is probably due to theirs large versatility in many different fields and conditions, resulting theirs properties either thermomechanic, thermophysic or thermodynamic, widely modellizable to specific purposes. The components selection' and their mixing process can be recognized as a true "solvent engineering project", being the final properties featured as a sum of contributions due to each component, and some residual contributions (actually, not at all predictable *a priori* on the basis of the current and acquired knowledge) neither in sign nor in intensity due to specific intermolecular interactions between different species.

Lacking of the general laws for mixed liquid systems, many different approaches for the analytical representation of these properties as a function of composition have been described in literature, both of empirical and theoretical nature. As a general rule, validation of the selected methods for analytical description of experimental values and for the recognition ability in correspondence of data gaps, derives from the fact that the average and maxima deviations for the correlated data are near to the average deviations quoted in experimental databases.

In summary, it is possible to assert that the correlative relationships and models presented in this paper may accurately predict each of the investigated property for the selected three-component mixtures on the whole temperature and composition ranges. As far as the excess mixing properties (or deviation functions) concerns (both for experimental data such as Δn, Δn, and/or derived quantities such as V^E, ΔG^{*E}, P^E, F_{μ}^{E}, R^E, ecc.), strong deviations from linear behaviour represented by a generic additive rule are always observed at all experimental conditions. The non-ideality is mainly associated with the presence of polar groups, such as –OH and $-OCH_3$, in the three components, which are responsible for the very effective heterocooperative hydrogen-bonding between different species. To gain further enlightnening information, for some additive properties with extensive units we have obtained the relevant partial excess molar properties for each *i*-th component, such as $\overline{V}_i^E$, $\overline{P}_i^E$, $\overline{F}_{\mu,i}^E$. As a rule, the most stressed species by mixing process is always DME, at each experimental condition and whichever property is taken into account, while ED and ME are almost comparatively perturbed by the presence of the cosolvents.

However, quite a remarkable experimental evidence coming out from the present work is that specific intermolecular interactions are very effective at binary level, but they seem to be more and more sleeping out in the three component mixtures, singular points as maximum deviations in excess properties being generally absent over the whole composition domain.

ACKNOWLEDGEMENTS

The authors are very grateful to all those research workers that, for conciseness, have not been cited in this paper. This is not a forgetfulness: their results and seminal contributions, their insight and inspiration of very high standard quality, generally have broadened our appreciation and understanding of the issues discussed. Furthermore, we thankfully acknowledge the many graduate students whose experimental work and contributions made over the last two decades have been capitalized for the drawing up of this paper.

Finally, we recognize the advice preciousness of the anonymous reviewers and colleagues around the world whose constructive criticisms have improved our works.

The MURST of Italy is gratefully acknowledged for the financial support.

REFERENCES

[1] Franks, F.; Ives, D.J.G. *Quart. Rev., Chem. Soc.* 1966, 20, 1-45.

[2] Franchini, G.C.; Tassi, L.; Tosi, G. *J. Chem. Soc. Faraday Trans. 1*,1987, 83, 3129-3138.

[3] Franchini, G.C.; Marchetti, A.; Tassi, L.; Tosi, G. *Anal. Chem.* 1990, 62, 1004-1010.

[4] Corradini, F.; Marcheselli, L.; Tassi, L.; Tosi, G.; Fanali, S. *Can. J. Chem.* 1993, 71, 1265-1272.

[5] Tassi, L. *J. Chem. Soc. Faraday Trans.*, 1993, 89, 733-738.

[6] Corradini, F.; Marchetti, A.; Tagliazucchi, M.; Tassi, L.; Tosi, G. *Can. J. Chem.*, 1994, 72, 339-344.

[7] Momchilov, A.; Manev, V.; Nasalevska, A.; Pasquali, M.; Pistoia, G. *J. Appl. Electrochem.* 1990, 20, 763–769.

[8] Pistoia, G. (Ed.), Lithium Batteries; Elsevier: Amsterdam (NL), 1994.

[9] Davis, M.I. *Chem. Soc. Rev.* 1993, 22, 127–134.

[10] Kay, R.L.; Vidulich, G.A.; Pribadi, K.S. *J. Phys. Chem.* 1969, 73, 445-447.

[11] Das, B.; Hazra, D. K. *J. Solution Chem.* 1998, 27, 1021-1031.

[12] Barthel, J.; Neueder, R.; Roch, H. *J. Chem. Eng. Data* 2000, 45, 1007-1011.

[13] Kinart, C.M.; Kinart, W.J.; Wiklinska, A. C. *J. Chem. Eng. Data* 2002, 47, 76-78.

[14] Cocchi, M.; Marchetti, A.; Pigani, L.; Sanna, G.; Tassi, L.; Ulrici, A.; Vaccari, G.; Zanardi, C. *Fluid Phase Equil.* 2000, 172, 93-104.

[15] Cocchi, M.; Marchetti, A.; Sanna, G.; Tassi, L.; Ulrici, A.; Vaccari, G. *Fluid Phase Equil.* 1999, 157, 317-342.

[16] Foca, G.; Manfredini, M.; Manzini, D.; Marchetti, A.; Pigani, L.; Sighinolfi, S.; Tassi, L.; Ulrici, A. *Intl. J. Thermophysics* 2004, 25, 839-855.

[17] Baraldi, P.; Giorgini, M.G.; Manzini, D.; Marchetti, A.; Tassi, L. *J. Solution Chem.* 2002, 31, 873-893.

[18] Marchetti, A.; Tagliazucchi, M.; Tassi, L.; Tosi, G. *J. Chem. Eng. Data* 1991, 36, 368-371.

[19] Franchini, G.C.; Marchetti, A.; Preti, C.; Tassi, L. *Ann. Chim. (Rome)* 1996, 86, 357-367.

[20] Corradini, F.; Marchetti, A.; Tagliazucchi, M.; Tassi, L.; Tosi, G. *Aust. J. Chem.* 1994, 47, 415-422.

[21] Baraldi, P.; Franchini, G.C.; Marchetti, A.; Sanna, G.; Tassi, L.; Ulrici, A.; Vaccari, G. *J. Solution Chem.* 2000, 29, 489-504.

[22] Pedrosa, G. C.; Salas, J. A.; Katz, M. *Thermochim. Acta* 1990, 160, 243-252.

[23] Viti, V.; Zampetti, P. *Chem. Phys.* 1973, 2, 233-237.

[24] Kuhn, L.P.; Wires, A.R. *J. Am. Chem. Soc.* 1964, 86, 2161-2165.

[25] Caminati, W.; Corbelli, G. *J. Mol. Spectrosc.* 1981, 90, 572-578.

[26] Straatsma, T.P.; McCammon, J. A. *J. Chem. Phys.* 1989, 90, 3300-3304.

[27] Buckley, P.; Brochu, M. *Can. J. Chem.* 1972, 50, 1149–1156.

[28] Ramanamurti, M.V.; Prabhu, P.V.S.S.; Bahadur, S.L. *Bull. Chem. Soc. Jpn.*1986, 59, 2341-2346.

[29] Eyring, H.; John, M.S. *Significant Liquid Structure*, Wiley, New York, 1969.

[30] Koga, Y.; Westh, P. *Bull. Chem. Soc. Jpn* 1996, 69, 1505–1508.

[31] Corradini, F.; Franchini, G.C.; Marchetti, A.; Tagliazucchi, M.; Tassi, L.; Tosi, G. *J. Solution Chem.* 1993, 22, 1019-1028.

[32] Corradini, F.; Marchetti, A.; Tagliazucchi, M.; Tassi, L.; Tosi, G. *Ann. Chim. (Rome)* 1994, 84, 397-415.

[33] Corradini, F.; Franchini, G.C.; Marchetti, A.; Tagliazucchi, M.; Tassi, L. *Can. J. Chem. Eng.* 1997, 75, 494-501.

[34] Partington, J. R. *An Advanced Treatise on Physical Chemistry*, Longmans Green, London, 1951.

[35] Monnery, W. D.; Svrcek, W.Y.; Mehrotra, A.K. *Can. J. Chem. Eng.* 1995, 73, 3 -40.

[36] Tamman, G.; Hesse, W. *Z. Anorg. Allg. Chem.* 1926, 156, 245-250.

[37] Williams, N.L.; Landel, R.F.; Ferry, J.D. *J. Am. Chem. Soc.* 1955, 77, 3701-3707.

[38] Oswal, S. *Can. J. Chem.* 1988, 66, 111-116.

[39] Pal, A.; Sharma, S. *J. Chem. Eng. Data* 1999, 44, 212-215.

[40] Cocchi, M.; Manfredini, M.; Marchetti, A.; Sighinolfi, S.; Tassi, L.; Ulrici, A.; Vignali, M. *Phys. Chem. Liq.* 2001, 39, 481-498.

[41] Marcus, Y. *J. Solution Chem.*, 1992, 21, 1217-1230.

[42] Eyring, G.; Powell, R.E.; Roseweare, W.E. *Ind. Eng. Chem.* 1941, 33, 837-841.

[43] Glasstone, S.; Laidler, J. K.; Eyring, H. *The Theory of Rate Process*, McGraw-Hill, New York, 1941.

[44] Corradini, F.; Marchetti, A.; Tagliazucchi, M.; Tassi, L.; Tosi, G. *Aust. J. Chem.* 1994, 47, 1117-1126.

[45] Read, T. M.; Taylor, T. E. *J. Phys. Chem.* 1959, 63, 58-67.

[46] Meyer, R.; Meyer, M.; Metzger, J.; Peneloux, A. *J. Chim. Phys. Phys-Chim. Biol.* 1971, 68, 406-412.

[47] Pimentel, C.G.; McClellan, L.A. *The Hydrogen Bond*, W.H. Freeman, London, 1960.

[48] Lange, E.; Robinson, A. L. *J. Am. Chem. Soc.* 1930, 52, 2811-2813.

[49] Akerlof, G. *J. Am. Chem. Soc.* 1932, 54, 4125- 4139.

[50] Franchini, G.C.; Marchetti, A.; Tagliazucchi, M.; Tassi, L.; Tosi, G. *J. Chem. Soc. Faraday Trans.* 1991, 87, 2583-2588.

[51] Goldoni, G.; Maffei, E.; Marcheselli, L.; Tassi, L. *J. Chem. Eng. Data* 1992, 37, 262-264.

[52] Fanali, S.; Goldoni, G.; Marcheselli, L.; Pistoni, G.; Tassi, L. *J. Chem. Soc. Faraday Trans.* 1992, 88, 2003-2006.

[53] Akhadov, Y.Y. Dielectric Properties of Binary Solutions – A Data Handbook, Pergamon Press, Oxford, 1981.

[54] King, M.; Queen, N. *J. Chem. Eng. Data*, 1979, 24, 178-181.

[55] Corradini, F.; Marcheselli, L.; Marchetti, A.; Tagliazucchi, M.; Tassi, L.; Tosi, G. *Aust. J. Chem.* 1993, 46, 1545-1555.

[56] Winkelmann, J.; Quitzsch, K. *Z. Phys. Chem.* (*Leipzig*) 1976, 257, 678-688.

[57] Winkelmann, J.; Quitzsch, K. *Z. Phys. Chem.* (*Leipzig*) 1976, 257, 746-756.

[58] Davis, M. I.; Douhéret, G. *Thermochim. Acta* 1986, 104, 203 - 222.

[59] Davis, M. I.; Douhéret, G. *Thermochim. Acta* 1987, 113, 369-377.

[60] Davis, M. I.; Douhéret, G. *Thermochim. Acta* 1987, 116, 183-194.

[61] Fialkov, Yu. Ya. *Zh. Fiz. Khim.* 1963, 37, 1757-1762.

[62] Fort, R. J.; Moore, W. R. *Trans. Faraday Soc.* 1966, 62, 1112-1119.

[63] Payne, R.; Theodorou, I. *J. Phys. Chem.* 1972, 76, 2892-2900.

[64] Lorentz, H. A. *Theory of Electrons*, Leipzig, 1906.

[65] Brocos, P.; Piñeiro, Á.; Bravo, R.; Amigo, A. *Phys. Chem. Chem. Phys.* 2003, 5, 550 – 557.

[66] Arago, D. F. J.; Biot, J. B. *Mem. Acad. Fr.* 1806, 7-11.

[67] Gladstone, J. H.; Dale, T. P. *Phil. Trans. R. Soc. London* 1863, 153, 317-343.

[68] Kurtz, S. S.; Ward, A. L. *J. Franklin Inst.* 1936, 222, 563-592.

[69] Oster, G. *Chem. Rev.* 1948, 43, 319-365.

[70] Eykman, J. F. *Ber. Deut. Chem. Ges.* 1906, 25, 3069-3079.

[71] Hawrylak, B.; Andrecyk, S.; Gabriel, C.E.; Gracie, K.; Palepu, R. *J. Solution Chem.* 1998, 27, 827-841.

[72] Fialkov, Yu. Ya.; Fenerli, G. N. *Russ. J. Inorg. Chem.* 1964, 9, 1205-1208.

[73] Wiener, O. *Ber. Schs. Ges. Wiss.* 1910, 62, 256-277.

[74] Heller, W. *Phys. Rev.* 1945, 68, 5-10.

[75] Parfitt, G. D.; Wood, J. A. *Trans. Faraday Soc.* 1968, 64, 805-814.

[76] Azim, A.; Munk, P. *J. Phys. Chem.* 1987, 91, 3910-3914.

[77] Azim, A.; Cheng, W.; El-Hibri, M. J.; Munk, P. *J. Phys. Chem.* 1988, 92, 2663-2668.

[78] Li, W. B.; Segré, P. M.; Gammon, R. W.; Sengers, J. V.; Lamvik, M. *J. Chem. Phys.* 1994, 101, 5058-5069.

[79] Franchini, G.C.; Marchetti, A.; Tassi, L.; Vaccari, G. *J. Chim. Phys.* 1998, 95, 1929-1950.

[80] Manfredini, M.; Marchetti, A.; Sighinolfi, S.; Tassi, L.; Ulrici, A. *J. Mol. Liq.* 2003, 102, 53-81.

In: Solution Chemistry Research Progress
Editor: Darian V. Bostrelli, pp. 51-68 ISBN 978-1-60021-993-1

Chapter 2

PROPERTIES OF INDIVIDUAL IONS IN SOLUTION

Yizhak Marcus
Hebrew University of Jerusalem, Israel

ABSTRACT

Knowledge of the physico-chemical properties of ions in solution is indispensable for understanding their behavior in laboratory, physiological, geological, and industrial environments. Apart from the transference numbers, conductivities and (self) diffusion coefficients that can be measured for individual ions in solution, most other desired properties can only be estimated properly by splitting measured values for electrolytes into the contributions of their constituent ions.

This pertains to such thermodynamic quantities as standard partial molar volumes, heat capacities, and entropies. Also of importance in this context are the standard molar Gibbs energy, enthalpy, and entropy of solvation and of transfer between solvents. The problem has so far found various solutions for obtaining individual ionic properties in selected solvents, foremost water, and to a lesser extent also in binary solvent mixtures.

The desiderata for methods for splitting the electrolyte data are stated, and the merits and disadvantages of existing methods are discussed in detail. In view of these, the requirements of future research in this direction are presented.

INTRODUCTION

Life on earth started in the oceans, as is generally accepted, and the oceans are huge masses of aqueous salt solutions, that is, of aqueous electrolytes. Electrolytes are salts, acids, or bases that dissociate at least partly into ions in solutions. Aqueous electrolytes also constitute a major and vital part of the bodies of the readers of this chapter. Much of the material world with which they deal results from chemical and metallurgical industries that involve processes carried out on electrolyte solutions. The food they eat is related indirectly to the fertilizer industry, some of the metals they use results from hydrometallurgical processes. The electricity they consume may be obtained from nuclear energy sources that involve processing of aqueous electrolytes. The readers of this chapter are likely to be active in chemical laboratory work, where solutions of electrolytes in water, non-aqueous solvents, and mixtures of these are commonly employed. Even synthetic organic processes often involve ionic species or charged transition states in solution. Biochemical processes involving enzymes take place in aqueous solutions and may have electric charges taking part in the interactions. Ions in solution are, therefore, ubiquitous constituents of the experience of the readers of this chapter.

In order to be proficient in the handling of ionic solutions and to make efficient use of them it is necessary to know their properties. For a given electrolyte and solvent these properties depend on the ambience and on the electrolyte concentration. The former involves the temperature T and the pressure P, and these may range from the extreme values in deep geological environments or in supercritical fluids that may be employed in industrial or laboratory processes to the mild ambience of the standard values specified for reporting thermodynamic quantities, i.e., 298.15 K (25 °C) and 0.1 MPa (= 1 bar = 0.98692 atm). The concentration, amount of substance per unit volume, generally expressed in molarity, c/mol dm^{-3}, ranges from infinite dilution to saturation. Beyond very dilute solutions it affects the ion-ion interactions and may affect the extent of the ionic dissociation (or its association). It is, therefore, necessary to specify the conditions under which the properties of the electrolytes are to be measured and discussed.

There are many kinds of such properties that are of importance and hence having been reliably measured and critically evaluated, or, on the other hand, not yet determined accurately nor available in the open literature. Due to the extremely large number of possible electrolyte/solvent systems that are, or could in the future be, employed, it is futile to expect to find already published the needed information for any specific electrolyte/solvent/property combination. As a first step to obtaining the required data for a given system, recourse is taken to theoretical considerations concerning the interactions occurring between the ions and the solvent (or solvents in a mixture) and other ions present.

Ionic Properties Pertaining to Ion-ion Interactions

Theoretical considerations regarding *ion-ion* interactions generally involve electrolytes as entire entities and do not concern individual ions. The well known Debye-Hückel theory [1] of very dilute electrolyte solutions and its limiting laws, pertaining to extreme dilution, as well as its extensions to higher concentrations but still dilute [2], apply irrespective of the natures of the electrolyte and the solvent. Such theories require the following properties of the electrolyte: the charge type (e.g., 1:1, 1:2, 2:3, etc. for the charges of the cation and anion) and a mean size of the ions (dispensable for the limiting laws). They require for the solvent its relative permittivity (dielectric constant), regarding the solvent as a continuum devoid of molecular structure (the so-called 'primitive model'). From these theories are obtained the primary quantities, the activity coefficient of the electrolyte and the osmotic coefficient of the solvent, related to the chemical potentials, i. e., the partial molar Gibbs energies of the electrolyte and solvent. Derived quantities, such as limiting laws of partial molar volumes, enthalpies, entropies, heat capacities, etc. are also obtained from such theoretical approaches.

Only rarely have efforts been made, such as Kieland's [3], to estimate individual ionic properties for activity coefficients, and such data have not found appreciable subsequent use, except in regards to pH measurements and their interpretation. The latter subject, though of great importance, is outside the scope of this chapter.

What is absent in the theoretical treatments of ion-ion interactions is an ability to *predict* the values of activity coefficients – as distinct from their *modeling*, e.g., by means of Pitzer's equations [4] – at concentrations beyond the validity of the simple extension of the Debye-Hückel theory. Such predictions must abandon the primitive model and take into account the molecular nature of the solvent as well as the properties of the individual ions. Such properties include besides the ionic size, shape, and polarizability, that affect the structure of the surrounding solvent, also the ability of the ion to interact with solvent molecules directly via electron pair and hydrogen bond donation and acceptance.

As an illustration, even a simple prediction of the density ρ of moderately concentrated aqueous salt solutions is beyond the capabilities of present-day theoretical treatments. Empirical expressions for such densities exist:

$$\rho = \rho^* + a(T)c + b(T)c^{3/2} \tag{1}$$

where ρ^* is the density of the solvent, c is the molar concentration and a and b are listed temperature dependent coefficients. The apparent molar volume of the salt ${}^{\varphi}V_2$ is related to these empirical coefficients on the one hand:

$${}^{\varphi}V_2 = [M_2 - 1000\{a(T) - b(T)c^{1/2}\}]/\rho^* \tag{2}$$

where M_2 is the molar mass (g mol^{-1}) of the salt, and to the theoretical expression:

$${}^{\varphi}V_2 = {}^{\varphi}V_2^{\infty} + S_V^{DH}c^{1/2} + b_Vc \tag{3}$$

on the other hand. Here ${}^{\varphi}V_2^{\infty}$ is the limiting value of the apparent and partial molar volume at infinite dilution, i.e., the standard partial molar volume, and S_V^{DH} is the limiting theoretical Debye-Hückel slope. However, there exists no theory that is able to predict the coefficient of the linear term, b_V, and to derive from it the descriptive coefficients a and b [5]. This, and similar quantities for other properties of moderately concentrated electrolyte solutions – not to speak of highly concentrated ones – require knowledge of individual ionic properties in solution, a challenge for further studies.

INDIVIDUAL IONIC PROPERTIES – ARE THEY MEASURABLE?

Some thirty years ago Conway [6] critically analyzed the methods then available for the estimation of the quantities pertaining to individual ions in solution. He delineated several areas in physical chemistry where such data would be useful if not essential. These include the specificities of ion-solvent interactions, effects of overlapping solvation spheres at moderate concentrations, specific effects in biophysics and ion binding in ion exchangers, membranes, polyelectrolytes, micelles, proteins, etc., individual ion effects at interfaces and electric double layers, transfer of ions between immiscible solvents, kinetic electrode effects, and some more.

Many of the properties of individual ions dealt with were thermodynamic, and this in spite of the firm tenet of chemical thermodynamics that only quantities pertaining to components that can be added independently to a solution (i.e., electrolytes or neutral combinations of cations and anions) have physical meaning and can be measured [7]. The addition of particles having charges of a single kind (e.g., positive ions) to a liquid produces a large space charge in the solution that depends on the external environment, shape, and dimensions of the solution vessel and has no chemical significance. Therefore, in the framework of classical chemical thermodynamics, the properties of individual ions had been considered to be meaningless, because they cannot be measured. On the other hand, statistical thermodynamics permits such properties as the solvation Gibbs energy of individual ions and derived quantities to be reliably calculated, as exemplified in the computer simulations referred to below [8, 9].

There are certain measurements and computations that can indeed be performed on ionic solutions and that pertain to individual ions. These involve structural, spectroscopic, and transport properties, but not thermodynamic ones in the strict sense of the term. Computer simulations by means of the Monte Carlo (MC) and molecular dynamics (MD) methods [8] provide structural information for individual ions, and the latter also provide certain spectroscopic and transport information. The primary result of the application of such computations, carried out on, say, one ion and 256 to 1024 solvent molecules in a 'box' defined by the ambience (temperature or energy and pressure), is the pair correlation function $g_{i,S}(r)$. This specifies the probability of finding the center of a solvent molecule S in a volume element at a distance r from the center of the ion i. This function has several peaks as the distance r increases, the first one being the most significant, pertaining to the mean distance $d_{i,S}$ between the ion and the center of the nearest solvent molecule. Other properties of the ion/solvent system, such as coordination numbers, can also be deduced from this function.

Structural information is obtained also from diffraction measurements on electrolyte solutions, such as x-ray and neutron diffraction. These data provide the mean distance d between the centers of adjacent ion and solvent molecules as well as the coordination number of the solvent around the ion [10]. The isotopic difference method introduced by Enderby et al. [11] to neutron diffraction measurements was an important advance enabling reliable information on certain ion/solvent systems to be obtained. From this mean distance $d_{i,S}$ ionic radii r_i in solution can be deduced, if the radius of the solvent molecule is defined and known. In the case of water as the solvent, the molecules of which have a collision diameter of 0.276 nm, hence a radius of 0.138 nm, the resulting ionic radii are [10]:

$$r_i = d_{i\text{-water}} - 0.138 \text{ nm} \tag{4}$$

Such radii in aqueous solutions correspond well with those obtained by Pauling from crystal structures, provided the radius of the water molecule is taken as an invariant quantity, but this view has been challenged [12]. Some more recent methods, such as EXAFS [13], have also been employed for the study of the structure of electrolyte solutions and applied to the properties of individual ions in them.

Diffraction measurements usually require an appreciable concentration of the electrolyte in the solution, say from 0.5 mol dm^{-3} and upwards. Both computer simulation and diffraction measurements can profitably be applied only for rather simple ions and solvent molecules with fairly few atoms in each (<10). For diffraction measurements this is due to complications arising from intra-molecular distances within solvent molecules and complex ions that cannot be readily disentangled from the overall ion-solvent distances. For computer simulations the problem is the incomplete knowledge of the potentials between the individual combinations of nuclei and electronic shells of the interacting ions and atoms in solvent molecules that is needed for the computations. Still, much progress has been made in this field to enable more complicated systems to be studied [9].

Spectroscopic measurements for the characterization of ionic properties may involve UV-vis, infrared, Raman, and NMR spectra. The electronic transitions of transition metal and f-block metal ions are often sensitive to the solvent in their coordination sphere and ligand-to-ion charge transfer also leads to measurable spectral changes in the UV-vis region. The spectral bands for certain bonds in solvent molecules (e.g., C=O, P=O, C≡N) in the infrared [14] and Raman [15] spectra of electrolyte solutions can be interpreted in terms of the effects of individual ions on the solvent molecules. However, it is rare to be able to find a new spectral band that is due to direct ion/solvent interactions. The NMR signal from an appropriate isotope of an atom in the ion or solvent molecule to be studied can be used to probe the environment of an ion in solution. Such data provide insight into the nature and extent of the coordination sphere as well as the dynamics of solvent exchange between this sphere and the bulk solvent [16]. These methods for obtaining properties of individual ions are not further discussed in this chapter.

Transport property measurements on electrolyte solutions are another way for obtaining some properties of individual ions. Tracer self-diffusion of isotopically labeled ions can be measured and provides data on their self-diffusion coefficients, D_i. These can be related to the mobility of the ion in an electrical field. The mobility u_i of the ion in such a field is its velocity under a unit electric potential gradient, which is proportional to the ionic conductivity λ_i:

$$u_i = \lambda_i/F \tag{5}$$

where F is faraday's constant. The limiting values at infinite dilution are according to the Nernst-Hartley equation [17]:

$$D_i^\infty = (RT/F^2)\,|z_i|^{-1}\,\lambda_i^\infty \tag{6}$$

Here D_i^∞ is the limiting self-diffusion coefficient of the ion i, R is the gas constant, z_i is the charge of the ion i, and λ_i^∞ is the limiting molar electrical conductivity of this ion. The latter quantity, on the other hand, can be reliably obtained as:

$$\lambda_i^\infty = t_i^\infty \Lambda^\infty \tag{7}$$

from the measurable transference number t_i^∞ of the i-ions extrapolated to infinite dilution and the limiting equivalent conductivity Λ^∞ of the electrolyte for which t_i has been measured. This quantity requires extrapolation of molar conductivities Λ measured in dilute solutions to infinite dilution via an expression that takes into account the electrophoretic and ionic atmosphere relaxation effects [18].

The electric conductivity of an ion depends on its size and on the viscosity of the solvent. The ion can, therefore, be assigned a so-called Stokes radius:

$$r_{i\,St} = (F^2/6\pi N_A)\,|z_i|/\eta^*\,\lambda_i^\infty \tag{8}$$

where N_A is Avogadro's number and η^* is the viscosity of the (neat) solvent. This radius may be related to the ionic radius r_i discussed above [19], but takes into account in some manner the solvation of the ion. It is the solvated ion that moves in the solution but it may leave some of its solvation shell behind as it moves: the electrophoretic effect. However, the quantities obtained from these transport measurements are not simply related to the static properties of ions in solution that are, in appropriate combinations of cations and anions, relevant to the thermodynamics of electrolyte solutions.

Nevertheless, there are two *chemical thermodynamic* quantities that have been claimed to be measurable for individual ions: molar volumes V_i and chemical potentials μ_i. The following discussion and the quoted numerical values pertain to infinite dilution of the electrolyte, T = 298.15 K, and ambient pressure, unless otherwise noted.

Zana and Yeager [20] obtained the standard partial molar volumes V^∞ of individual ions by means of measured ultrasonic vibration potentials, Φ_0, of dilute electrolyte solutions. The relevant equations with the unknown V^∞_+ for the cation of a 1:1 electrolyte and V^∞_- for its anion are:

$$\Phi_0 = C[t_+(M_+ - \rho^* V^\infty_+) - t_-(M_- - \rho^* V^\infty_-)] \tag{9}$$

$$V^\infty = V^\infty_+ + V^\infty_- \tag{10}$$

Here C is a known constant depending on the solvent (here water) and the temperature only, t as above is the experimentally measured transference number and M is the molar mass of the designated ion, and ρ^* is the density of the solvent. Simultaneous solution of the two equations then yields the two $V^\infty_\pm$ (V^∞_+ and V^∞_-) values. However, as Conway [6] and Marcus and Hefter [21] pointed out, eq. (9) deals with a hydrodynamic transport quantity and eq. (10) with a thermodynamic one, and the implicit assumption that the $V^\infty_\pm$ of eq. (9) and (10) pertain to the same quantity cannot be proved. Still, the resulting value for the aqueous hydrogen ion, $V^\infty(H^+, aq) = -5.4\pm2.0\ cm^3\ mol^{-1}$ at 22 °C, is in good agreement with other estimated values (see below), but has a large uncertainty due to its derivation as an average of results from many electrolytes, including those with multiply charged ions.

The individual ionic standard chemical potentials μ^∞_i are treated in the following discussion of their solvation properties, involved in ion-solvent interactions.

IONIC PROPERTIES INVOLVED IN ION-SOLVENT INTERACTIONS

Theoretical considerations of *ion-solvent* interactions necessarily devolve around individual ions. When infinitely dilute solutions of an electrolyte are dealt with, each cation and anion is surrounded solely by solvent molecules and ion-ion interactions can be disregarded. The properties deduced for the individual ions are then *additive*, and values for the entire electrolyte – that are measurable in principle – can be reconstituted from the individual ionic properties, weighted according to the respective stoichiometric coefficients. This treatment can then be applied to *predict* values for electrolytes, for which no value of the property has been (or can be) measured directly. The procedure is in practice turned around, so that measured values for electrolytes, extrapolated appropriately to infinite dilution, are split into the contributions of the individual ions. Appropriate extrapolation involves the theories of ion-ion interactions mentioned above, but the splitting into single ionic contributions is a major concern of this chapter.

The so-called 'real' individual ionic electrochemical potential μ_i^R can be determined for an ion from EMF data by the vertical jet method [22]. An air gap separates a solution jet flowing down the center of a vertical tube from another (reference) solution flowing down the interior surface of the tube, both solutions being half-cells of a galvanic cell. This quantity, extrapolated to infinite dilution, $\mu_i^{R\infty}$, then leads to the ionic standard chemical potential μ_i^{∞} by means of the surface potential difference $\Delta\chi$ between the gas (air) and the jet solution solvent:

$$\mu_i^{R\infty} = \mu_i^{\infty} + zF\,\Delta\chi \tag{11}$$

Here z is taken in the algebraic sense. However, so far $\Delta\chi$ can only be estimated but not be experimentally determined reliably. The surface potential arises from the distribution of solvent dipoles (solvents of electrolytes are generally polar) at the surface and the electrical double layer due to ions adsorbed at the surface [23]. In the case of water as the solvent, values within the wide range $-1.10 \le \Delta\chi/\mathrm{V} \le +0.50$ have been reported, with +0.10 V being recently estimated as the most nearly correct [24]. The coupling between the two terms on the right hand side of eq. (11) and the probable error of ~ ±0.05 V in $\Delta\chi$ that was estimated causes a probable error of ~ ±5 kJ mol^{-1} in the resulting μ_i^{∞} for a univalent ion.

The cluster pair approximation of Tuttle et al. [25] served as the basis for the recent calculation by Kelly et al. [26] of the hydration and solvation Gibbs energies of the hydrogen ion, hence the absolute potential of the normal hydrogen electrode. The following expression couples these quantities for aqueous solutions:

$$E°[\mathrm{H^+(aq)/H_2(g)}] = [\Delta_f G°(\mathrm{H^+(g)}) + \Delta_{hydr}G°(\mathrm{H^+})]/F \tag{12}$$

The standard molar Gibbs energy of formation of the gaseous hydrogen ion $\Delta_f G°(\mathrm{H^+(g)})$ can be calculated unambiguously from experimental data, using the measured dissociation energy of $H_2(g)$ and the ionization potential of $\mathrm{H(g)\rightarrow H^+(g)}$. However, the standard molar Gibbs energy of hydration of the hydrogen ion $\Delta_{hydr}G°(\mathrm{H^+})$ must be obtained by other means. The operative expression (for water as the solvent) used by Kelly et al. [26] is:

$$\begin{aligned}&\tfrac{1}{2}[\Delta_{hydr}G^{*con}(\mathrm{C^+}) - \Delta_{hydr}G^{*con}(\mathrm{A^-})] + \tfrac{1}{2}[\Delta_{clust}G°_n(\mathrm{C^+}) - \Delta_{clust}G°_n(\mathrm{A^-})] =\\ &\tfrac{1}{2}[\Delta_{clust}G°_{n+1,\infty}(\mathrm{C^+}) - \Delta_{clust}G°_{n+1,\infty}(\mathrm{A^-})] + \Delta_{hydr}G^*(\mathrm{H^+})\end{aligned} \tag{13}$$

The conventional standard molar Gibbs energies of hydration in the first couple of terms on the left hand side, $\Delta_{hydr}G^{*con}(\mathrm{C^+})$ and $\Delta_{hydr}G^{*con}(\mathrm{A^-})$, are based on the convention that standard molar Gibbs energy of formation of the aqueous hydrogen ion is zero, $\Delta_f G^{*con}(\mathrm{H^+,aq}) = 0$. The * superscript for the standard state means that the same concentration, 1 mol dm^{-3}, is used for the gas and solution phases, ensuring that the derived quantities pertain solely to the interaction of the ion with its surrounding solvent and does not involve the compression of the gaseous species. The correction for the Gibbs energy from the usual standard state for the gaseous ion of 0.1 MPa, denoted by °, is 7.93 kJ mol^{-1} [27]. The conventional standard molar Gibbs energies of hydration are experimentally available for a large number of cations C^+ and anions A^-, since they are due to differences and sums between the values of species with charges +1 and −1, hence for neutral combinations that are thermodynamically unimpeachable. The second couple of terms on the left hand side are the sums of the stepwise clustering Gibbs energies of the cation and anion with $n \leq 6$ water molecules in the gas phase, also experimentally or computationally available quantities. The essence of the cluster pair approximation is to find pairs of cation C^+ and anion A^- for which the difference of the first couple of terms on the right hand side vanishes: $\Delta_{clust}G°_{n+1,\infty}(C^+) - \Delta_{clust}G°_{n+1,\infty}(A^-) = 0$. This procedure, then, leads to the value $\Delta_{hydr}G^*(H^+) = -1113 \pm 8$ kJ mol^{-1}. Insertion of this quantity into eq. (12) and taking $\Delta_f G°(H^+(g)) = 1516.9$ kJ mol^{-1} results in the standard potential of the normal hydrogen electrode $E°[H^+(aq)/H_2(g)] = 4.19 \pm 0.08$ V. Values for this quantity in some other solvents have also been derived by the same procedure when the clustering Gibbs energy data are known [26b].

In eq. (11) above the standard molar Gibbs energy of formation of the gaseous hydrogen ion can be subtracted from both sides. Then the left hand side of the resulting eq. (14):

$$\Delta_{hydr}G^{*R}(H^+) = \Delta_{hydr}G^*(H^+) + zF\,\Delta\chi \tag{14}$$

is the measured [22] real standard molar Gibbs energy of hydration, −1080±2 kJ mol^{-1}. From this and the value $\Delta_{hydr}G^*(H^+) = -1113 \pm 8$ kJ mol^{-1} the surface potential of water $\Delta\chi = 0.34 \pm 0.08$ V results, different from the previous estimate. The rather large probable error in these derived quantities, and the assumptions and approximations that are involved in these so-called 'direct' methods, should be noted.

This calculation of $\Delta_{hydr}G^*(H^+)$ was reported very recently, requiring data obtained also from quantum chemical computations of clustering Gibbs energies not feasible many years ago. However, the idea behind eq. (13) of finding a cation and an anion for which the unknown difference between certain thermodynamic values should vanish, is not new. Halliwell and Nyburg [28] employed this device more than 40 years ago in order to estimate the enthalpy of hydration of the hydrogen ion. The operative equation in this case is:

$$\tfrac{1}{2}[\Delta_{hydr}H^{*con}(C^+) - \Delta_{hydr}H^{*con}(A^-)] = \tfrac{1}{2}[\Delta_{hydr}H^*(C^+) - \Delta_{hydr}H^*(A^-)] - \Delta_{hydr}H^*(H^+) \tag{15}$$

The left hand side, again, is a measurable quantity, based on the convention that $\Delta_f H^{*con}(H^+,aq) = 0$, with data known for many cations and anions. The * standard state for the gaseous ion, 1 mol dm^{-1} differs from the usual one for the enthalpy of hydration of 0.1 MPa (~ 1 Atm), denoted by °, by 2.29 kJ mol^{-1} [27]. The two items in the square brackets on the right hand side should depend inversely on the sizes of the hydrated ions. Therefore a plot of the left hand side against the volumes of the hydrated ions, $(3/4\pi)[(r_{ion}/nm) + 0.138]^{-3}$, extrapolated to zero should then yield the desired value, that comes out as $\Delta_{hydr}H^*(H^+) = -1088 \pm 10$ kJ mol^{-1}. As with all the approximations employed for obtaining the individual ionic properties, this approach has also met objections, the extrapolation function, $[(r_{ion}/nm) + 0.138]^{-3} \rightarrow 0$, having been attacked.

STANDARD IONIC SOLVATION AND PARTIAL MOLAR QUANTITIES

As mentioned above, Conway [6] examined the methods known at the time for the estimation of standard molar thermodynamic quantities of single ions, rating their reliability from poor through fair and good to excellent. The starting points, as also described above, are the conventional values, based on the conventions for the Gibbs energy and enthalpy of formation of the aqueous hydrogen ions being zero, as are its standard partial molar entropy, heat capacity, and volume, and the normal electrode potential. Once the so-called absolute value for one ion are determined by means of an extra-thermodynamic assumption, those for all the other ions are derived from the additivity principle, operative at the standard state of infinite dilution. The conventional values have been compiled and critically examined in several publications, that of the present author [29] being an example. Such values may be refined somewhat in later publications when better experimental data have been obtained, but the changes are expected to be minor. The recommendations made by Conway concerning the standard molar Gibbs energy and enthalpy of hydration of the hydrogen ion were based on "assumptions and principles ... [that] are the least objectionable", and this seems to be the best that can be said about them and about some of those described in the following.

Somewhat different values then those recommended by Conway [6] for the standard molar enthalpy and Gibbs energy of hydration have subsequently been suggested by others. Those shown in [29] for the enthalpy are based on the TATB assumption [30], where TATB stands for tetraphenylarsonium tetraphenylborate:

$$\Delta_{\text{hydr}}H^\circ(\text{Ph}_4\text{As}^+) = \Delta_{\text{hydr}}H^\circ(\text{BPh}_4^-) = \Delta_{\text{hydr}}H^\circ(\text{Ph}_4\text{P}^+) \quad (16)$$

This assumption is based on the fact that the constituent ions of TATB are univalent, large, and similar in most respects except for the sign of the charge that is buried well within the hydrocarbon envelop (see also further below). Tetraphenylphosphonium ions can replace the tetraphenylarsonium ones for the purpose of this assumption (the TPTB assumption). In particular, this assumption makes sense for the enthalpy of hydration, since this is a small quantity (–47 kJ mol^{-1}) compared with values for small ions such as potassium (–334 kJ mol^{-1}) or chloride (–367 kJ mol^{-1}) ones, so that small discrepancies (10%) in the sizes and interactions of the TATB ions should not cause appreciable differences in their enthalpies of hydration [31]. According to the assumption (16) $\Delta_{\text{hydr}}H^\circ(\text{H}^+) = -1103\pm7$ kJ mol^{-1} and $\Delta_{\text{hydr}}H^*(\text{H}^+) = -1101\pm7$ kJ mol^{-1}. Note that the latter value is *less negative* than the value of $\Delta_{\text{hydr}}G^*(\text{H}^+)$ shown above as the result of the cluster pair approximation method [26a], so that their combination leads to a *positive* entropy of hydration that is contrary to any reasonable estimate of this quantity.

The standard molar entropy of the hydrogen ion $S^\circ(\text{H}^+,\text{aq})$ can be obtained from the potential of a mercury electrode at the point of zero charge or from thermocells. The assumption involved in the former method is that the temperature dependence of the surface potential of mercury in water is negligible. The corresponding assumption for the other method relates to the entropy of transport across a boundary of similar solutions at different temperatures [6, 19]. Consistent values are obtained from both methods: $S^\circ(\text{H}^+,\text{aq}) = -22.2\pm1.4$ J K^{-1} mol^{-1} [6], a value that has not been seriously challenged since. However, the more recent paper by Schmid et al. [32], suggesting a much smaller negative value, requires comment. It depends on gas-phase clustering entropies and an apparent similarity between the hydration of the hydrogen and the hydroxide ions. The adoption of the Schmid et al. value, $S°(H^+,aq) = -5.45$ J K^{-1} mol^{-1} would increase that of fluoride anions to 25.1 J K^{-1} mol^{-1}, which is much different from their hydroxide value $S°(OH^-,aq) = -5.45$ J K^{-1} mol^{-1}. On the other hand, on the basis of Conway's choice [6], the value of $S°(OH^-,aq) = 11.3$ J K^{-1} mol^{-1}, is near that of the fluoride anion, $S°(F^-,aq) = 8.4$ J K^{-1} mol^{-1}, which interacts with the water by hydrogen bonding in a manner very similar to the hydroxide one, and is to be preferred.

In fact, the TATB assumption was suggested by Cox and Parker [33] for obtaining this quantity, $S°(H^+,aq)$, analogous with eq. (16) with $S°$ replacing $\Delta_{hydr}H$. Recent determinations of the entropies and aqueous solubilities of crystalline tetraphenyl-arsonium and -phosponium perchlorates, yielded values of –40.9 and –33.4 J K^{-1} mol^{-1} for $S°(H^+,aq)$ from the TATB and the TPTB assumptions, respectively [34]. However, the standard molar entropies of aqueous TATB and TPTB have large values, 733 and 718 J K^{-1} mol^{-1}, so that any inaccuracy of the assumption of equal entropies of their ions has appreciable consequences. The values of $S°(H^+,aq) = -22.2$ J K^{-1} mol^{-1} is therefore still preferred.

With the standard molar entropy value for the gaseous hydrogen ion, obtained unequivocally by the Sackur-tetrode expression as $S°(H^+,g) = 108.9$ J K^{-1} mol^{-1}, the standard molar entropy of hydration of the hydrogen ion, $\Delta_{hydr}S°(H^+) = -131\pm1.4$ J K^{-1} mol^{-1} results. Conversion to the 1 mol dm^{-3} standard state for the gaseous hydrogen ion by subtraction of 18.9 J K^{-1} mol^{-1} [27] this then leads to $\Delta_{hydr}S^*(H^+) = -150$ J K^{-1} mol^{-1}. The Gibbs energy of hydration should therefore be *less negative* than the enthalpy of hydration by $T\Delta_{hydr}S^{\circ \text{ or } *} =$ 44.7 kJ mol^{-1}, a quantity that is incompatible with the estimates given above. In fact, Kelly et al. [26a] mentioned results from recent bulk water computer simulations that suggest a value for the absolute hydration Gibbs energy of the hydrogen ion that is close to –1054 kJ mol^{-1}, which is much better compatible with the entropy and enthalpy values of –150 J K^{-1} mol^{-1} and –1101 kJ mol^{-1} adjusted for the standard state of 1 mol dm^{-3} for the gaseous ions from those used by the present author [29] for the gaseous standard state of 0.1 MPa.

The individual standard molar ionic volumes in aqueous solutions have also been considered by Conway [6], who arrived at $V°(H^+,aq) = -4.7\pm0.5$ cm^3 mol^{-1} as his recommended value. Somewhat earlier Millero [35], whose compilation of the conventional values (i.e., $V^{\circ\,con}(H^+,aq) = 0$) has served for the basis of many attempts to obtain the absolute value, arrived at essentially the same value, –4.7±1.1 cm^3 mol^{-1}, as the mean of several more or less reliable estimates, ranging from –6.0 to –2.6 cm^3 mol^{-1}. This chosen value agrees with that obtained by Zana and Yeager [20] from vibration potentials, $V^{\infty}(H^+, aq) = -5.4\pm2.0$ cm^3 mol^{-1} (at 22 °C), see above. The analysis by Hefter and Marcus [36] of Milero's treatments [35] leads, however, to the choice of the ratio of the van der Waals volumes of the TATB ionic constituents as the best basis for an unequal splitting of $V^{\infty}(Ph_4As^+BPh_4^-)$ to yield $V^{\infty}(H^+, aq) = -5.5$ cm^3 mol^{-1}, the volume employed in subsequent publications too [29]. The difference with respect to the value by Millero is within the error limits ascribed to the latter and the value used [29] is nearer the Zana and Yeager one. This value was found more recently by Hedwig and Hakin [37] to be halfway between and within the ascribed uncertainties of the van der Waals volume ratio splitting of $V^{\infty}(Ph_4P^+BPh_4^-)$: –4.6 cm^3 mol^{-1}, and the value –6.6 cm^3 mol^{-1}, arising from the 2 cm^3 mol^{-1} difference between cation and anion of this salt in all solvents suggested by Hefter and Marcus [36]. The even splitting, i. e., the TATB or TPTB assumptions according to eq. (16), should *not* be invoked for the estimation of the individual ionic partial molar volumes, because real differences between the volumes of the cation and anion of these salts are known to exist.

A method for obtaining individual standard molar ionic volumes that has fairly often been employed in the past is the extrapolation to zero of values of $V^{\infty}(R_4NX, S)$ with respect of either the relative molar mass M of the cations [38] or the number N_C of carbon atoms of the alkyl chains R [39, 40]. This extrapolation should yield the standard partial molar volume of the anion X, $V^{\infty}(X^-, S)$, as the intercept. As pointed out by Hefter and Marcus [41], there is no logical argument for preferring the one or the other extrapolation, but the resulting values for $X^- = Br^-$ and S = water are different: 30.6 $cm^3\ mol^{-1}$ for extrapolation against M and 50.7 $cm^3\ mol^{-1}$ for extrapolation against N_C [41]. A similar procedure for obtaining individual ionic standard partial molar heat capacities, $C_P^{\infty}(R_4NBr, S)/J\ K^{-1}\ mol^{-1}$ at 298.15 K yields intercepts for $C_P^{\infty}(Br^-, S)/J\ K^{-1}\ mol^{-1}$ that are -303 ± 45 and -422 ± 49 respectively for the two extrapolations vs. M and vs. N_C in water and -43 ± 13 and -84 ± 13 in methanol, i.e., differing according to the abscissa used [19]. This discrepancy obviates the usefulness of such extrapolations for obtaining individual standard partial molar quantities of ions.

The individual standard molar ionic heat capacities at constant pressure in aqueous solutions have not fared as well as the entropies and volumes, since no reliable method for splitting the salt values into the contributions from their constituent ions has been found so far, and the extrapolation of the R_4NX values, as seen above, leads to equivocal values. Conway [6] quoted values of $C_p^{\infty}(H^+,aq)$ ranging from -61 to 142 $J\ K^{-1}\ mol^{-1}$. The value $C_p^{\infty}(H^+,aq) = -71\pm14\ J\ K^{-1}\ mol^{-1}$ was suggested by Abraham and Marcus [42] on the basis of the TPTB assumption, analogous to eq. (16) with C_p^{∞} replacing $\Delta_{hydr}H°$. The tetraphenylphosphonium salt was preferred to the -arsonium one because the size of the cation of the former is more nearly like that of the tetraphenylborate anion so that the heat capacities of the gaseous ions differ by only 0.7% for TPTB. The even split of $C_p^{\infty}(Ph_4P^+BPh_4^-)$ then yields aqueous values summing to within the probable errors of the experimental determinations. More recently Hedwig and Hakin [36] arrived at $C_p°(H^+,aq) = -62\ J\ K^{-1}\ mol^{-1}$ at 298.15 K from more accurate data for $C_p^{\infty}(Ph_4P^+BPh_4^-)$. As for the standard partial molar entropies, and contrary to the molar enthalpies of hydration, the use of the TPTB assumption is based on the splitting of a large value, 2282 $J\ K^{-1}\ mol^{-1}$, hence the accuracy of the result is still open to improvement. The compiled values [29] on this basis may, therefore, need revision if a better method for splitting the salt values is found in the future.

Several other individual ionic properties were discussed by Conway [6]: partial molar compressibilities, dielectric decrements, NMR chemical shifts, viscosity B-coefficients, and standard molar thermodynamic functions of transfer from one solvent to another. Only the latter two items are discussed here further, having since been dealt with by the present author and coworkers.

The viscosity B-coefficient is the slope of a plot of the left hand side of:

$$(\eta - \eta^*)/\eta^*\ c^{1/2} = A + B\ c^{1/2} \tag{17}$$

against the square root of the molar concentration of an electrolyte, $c^{1/2}$, where η is the viscosity of the solution and η^* that of the solvent. In the case of the B-coefficients, the guiding principle for splitting the salt values into the ionic contribution is the ionic mobility defined in eq. (5) or the Stokes radius defined in eq. (8). Again, attempts were made [6] to use an electrolyte, such as $Bu_4^+BPh_4^-$, the ions of which have similar but small mobilities:

$u(Bu_4N^+,aq) = 2.021$ and $u(BPh_4^-,aq) = 2.062$ (in 10^{-8} $m^2s^{-1}V^{-1}$). The mobilities are small due to the large ionic sizes, and differ by only 2%. Such an attempt may, however, cause large errors, since the *B*-coefficient for the aqueous $Bu_4^+BPh_4^-$ is large, 2.37 dm^3 mol^{-1} [43]. It is better to employ a salt, the ions of which have larger mobilities, but that has a small *B*-coefficient. The salt KCl having $B = -0.014$ dm^3 mol^{-1} was often used for this purpose following Gurney [44], with mobilities $u(K^+,aq) = 7.618$ and $u(Cl^-,aq) = 7.918 \times 10^{-8}$ m^2 s^{-1} V^{-1} at 298.15 K, differing by <4% over the temperature range 15 to 42 °C. The convention $B(K^+,aq) = B(Cl^-,aq)$ over this temperature range was thus established. However, other salts have more nearly equal mobilities of their ions over a wider temperature range, $u(Rb^+,aq) = 8.066$ and $u(Br^-,aq) = 8.097 \times 10^{-8}$ m^2 s^{-1} V^{-1} at 298.15 K, differing by <1% for the range 0 to 55 °C, and $B(Rb^+,aq) = B(Br^-,aq)$ was more recently proposed by Jenkins and Marcus [43]. With $B(RbBr,aq) = -0.066$ dm^3 mol^{-1}, still a small quantity, this choice would lead to small errors if the equal split between cation and anion is challenged for some reason. (This new scale differs from the KCl scale by only ±0.002 dm^3 mol^{-1} at 298.15 K for any ion.) Note that these scales pertain to aqueous solutions only, since in non-aqueous solvents the *B*-coefficients are considerably larger and no well established splitting of salt values into ionic contributions exists [43].

IONIC FUNCTIONS OF TRANSFER BETWEEN SOLVENTS

The following standard molar thermodynamic functions of transfer of ions from a source solvent (reference solvent) W to a target solvent S, $\Delta_t Y^\infty(\text{i, W}\rightarrow\text{S})$, have been treated, where *Y* stands for Gibbs energy, enthalpy, entropy, heat capacity, and volume. For all these quantities $\Delta_t Y^\infty(\text{i, W}\rightarrow\text{S})$ is the difference between the standard partial molar thermodynamic functions in W and S (for $Y = G$ this pertains to the chemical potentials). For the Gibbs energy and enthalpy the quantity $\Delta_t Y^\infty(\text{i, W}\rightarrow\text{S})$ can be construed as the difference between the standard molar Gibbs energy and enthalpy of solvation of the ion in W and in S. The latter quantities are much larger than the corresponding quantities for transfer between W and S, so if the latter can be determined directly and, say, the value for solvation in W (for hydration if W is taken as water) is known with good accuracy, the value of solvation in the solvent S can be obtained much more accurately than by independent estimates. The Gibbs energy of transfer has found the widest application, since it describes the preference of the ion for its solvent environment, whether W or S. Commonly, but not exclusively, W is taken as water and S may also be a solvent mixture, where preferential solvation of the ion may take place. Reviews of these quantities for transfer into neat non-aqueous solvents and their mixtures with water have been published by the present author and coworkers among several others: $Y = G$ [29, 45-53] and specifically $Y = G$ for cations [54] and for anions [55,56], $Y = H$ and S [57-59], $Y = C_p$ [60,61], and $Y = V$ [34, 40,62].

Several extra-thermodynamic assumptions have been proposed over the years to split the measurable electrolyte transfer data into their individual ionic contributions, but only three have been sufficiently widely employed to warrant discussion here. Objections were raised against each of them and there is no way to decide which is more nearly correct, only the objections to which are the least grave. Marcus [63] discussed the methods available at the time – none was seriously added since – and these were more recently again briefly discussed by Kalidas et al. [54] concerning the Gibbs energy of transfer (earlier discussion of this by Parker and Alexander [64] ought also to be mentioned), by Hefter et al. [59] concerning the enthalpy and entropy of transfer, and by Hefter and Marcus [41] concerning the partial molar volumes of ions in non-aqueous solvents.

The merits and drawbacks of using the 'real' electrochemical potential $\mu_i^{R\infty}$ for the splitting of the standard molar Gibbs energy of hydration of ions are discussed above. For the transfer of ions between W and S the difference in surface potentials, $\Delta_{W\to S}\Delta\chi$ must be known. Although these have been estimated, the derived values generally depend on the use of other methods for obtaining $\Delta_t G^\infty$(i, W→S). The recent work of Kelly et al. [26b], using the cluster pair approximation for estimation of the solvation Gibbs energy of ions and the normal hydrogen electrode potential in methanol, acetonitrile, and dimethylsulfoxide should be mentioned in this connection. Although the standard molar Gibbs energies of solvation of individual ions in these solvents are valuable quantities for themselves, the derivable Gibbs energies of transfer would be small differences between large numbers with considerable probable errors (~±8 kJ mol^{-1} each) and no warranty can be given for the resulting preferences of the ions between a pair of solvents.

The negligible liquid junction assumption for the determination of $\Delta_t G^\infty$(i, W→S) and its temperature derivative from EMF data has found many adherents. It requires a suitable salt bridge to be inserted between the solutions of an electrolyte in the two solvents W and S, in which electrodes responsive to the same ion i are placed. Optimal results have been obtained with 0.1 mol dm^{-3} of Et_4N^+ picrate in acetonitrile as the salt bridge electrolyte. The ions of this salt have nearly the same mobilities and are poorly solvated, so that the premises for a negligible liquid junction potential against any electrolyte solution are met [65]. A test of the applicability of this extrathermodynamic assumption is that similar EMFs should be obtained when different solvents are used in the bridge (although, as said, acetonitrile is optimal). However, cases were found where this does not work properly, mainly when the solvent W is water or when S is formamide, for no apparent reason. Therefore, the generality of the method was questioned.

Complex couples that consist of a neutral molecule and an ion, where both have very similar structures, are large enough so that the electrostatic effects of the solvation of the ion are small, and having the charge buried inside an inert 'cage' so that it is not specifically solvated by different solvents, have been widely employed for obtaining $\Delta_t Y^\infty$(i, W→S) values. The use of a cryptand for sequestering an alkali metal cation was first suggested by Villermaux and Delpuech [66], but their method needed the cumbersome use of solvent vapor pressure measurements. The method was later developed for the use of convenient electrochemical measurements and applied to $Y = G$, H, and S, as discussed in detail be Marcus [67]. The use of, e.g., the cryptand C222 (hexaoxa-1,10-diazabicyclo(8.8.8.)hexacosane), has found several proponents, the premises being that:

$$\Delta_t G^\infty(\mathrm{MC222}^+, \mathrm{W\to S}) = \Delta_t G^\infty(\mathrm{C222, W\to S}) \quad (18a)$$

and

$$\Delta_t H^\infty(\mathrm{MC222}^+, \mathrm{W\to S}) = \Delta_t H^\infty(\mathrm{C222, W\to S}) \quad (18b)$$

The values for the neutral ligand, C222, are obtainable experimentally from solubility and calorimetric measurements. A test of this approach is that the results should be independent of the nature of the counter-ion present (provided no ion pairing occurs) and was found to be valid within ±2 and ±6 kJ mol^{-1} for $Y = G$ and $Y = H$, respectively, when W and S are both dipolar aprotic solvents.

More widely employed were redox couples, in which a transition metal ion is enclosed by a ligand cage and the resulting complex has zero or unit charge. For such systems it is assumed that the redox couple can be employed as an invariant reference electrode, not sensitive to the nature of the solvent in which it is used. Earlier the ferrocene/ferricinium couple was used [68] but later the bis(biphenylchromium(0/I)) couple [69] was deemed to be more advantageous. Recently, Krishtalik et al. [70] re-proposed the cobaltocene/cobalticenium couple, in fact the two couples $CoCp_2^+/CoCp_2$ and $CoCp_2/CoCp_2^-$ (Cp = cyclopentadienyl), as having their own advantages. These approaches are based on voltammetric measurements, e.g., polarographic half-wave potentials, which are readily carried out accurately. The assumption of the independence of the electrode potential of such a couple from the solvent was, however, challenged, it being argued that the electrostatic effects of the solvation of the charged part of the couple were neglected. Still, such effects should be small, although the sizes of the complexes making up the couples are not very large, and the effects can be accounted for by use of the Born equation. This objection can be minimized if transfers between solvents of similar permittivities (say within 10 units in the relative permittivity ε_r) are considered, for which purpose acetonitrile, with $\varepsilon_r = 35.94$ is a good choice for the reference solvent W. More serious is the contention that the sandwich complexes are open in their equatorial surface, permitting specific solvation effects to occur. These extra-thermodynamic assumptions have invariably been applied to the transfer of cations, but the test that the same results, standard EMFs or half-wave potentials, are obtained irrespective of the accompanying anions present in the solution, has practically never been demonstrated to apply.

The recent suggestion [70] of comparison of the results from the two couples $CoCp_2^+/CoCp_2$ and $CoCp_2/CoCp_2^-$ leads naturally to the extra-thermodynamic assumption of using a reference electrolyte consisting of a cation and an anion that are univalent, are similar in all respects except for the sign of the charge, are large and nearly spherical, with the charges well buried inside, and having a more or less inert periphery. The ions of such an ideal electrolyte would have interactions with the surrounding solvent molecules similar to those of a neutral analog of similar structure, due to dispersion forces. For such ions, the ion-dipole electrostatic effects should be minimal and be the same for cation and anion, and only a small ion-quadrupole interaction with the solvent, differing between cation and anion, still occurs. TATB and TPTB, cf. eq. (16), have been suggested as most nearly approximating the desiderata for such a reference electrolyte. TPTB was first suggested by Grunwald et al. [71], who employed vapor pressure measurements, but this was later replaced by TATB by Alexander and Parker [72], using a more convenient solubility method. It is the latter reference salt that has by now gained very wide applications.

That TPTB and TATB are imperfect implementations of the reference electrolyte assumption was acknowledged by several workers. One objection to the even split of thermodynamic quantities between the cation and anion of these salts is their unequal sizes. Tetraphenylmethane is the best neutral analog of tetraphenylborate and tetraphenylsilane is the corresponding one for tetraphenylarsonium on account of the ionic and molecular sizes, the ratio of the van der Waals volumes being [36]:

$$V_{vdW}(Ph_4As^+)/V_{vdW}(BPh_4^-) = 1.0426 \quad (19)$$

The discrepancies between these ions regarding $\Delta_t G^\infty$ on account of the electrostatic and dispersion interactions are proportional to functions of the van der Waals radii: to r_{vdW}^{-1}, 2.4%, and to r_{vdW}^2, 1.2%, respectively [63]. The effect on $\Delta_t G^\infty$ of the different ion-quadrupole interactions with solvents was estimated by Kim [73], but the estimates involved several unverifiable assumptions and generally unknown solvent molecular quadrupole moments. Due to all these known effects, the even (50/50) splitting of $\Delta_t G^\infty$(TATB, W→S), analogous to eq. (16), was rejected by Kim in favor of a 53/47 splitting. The difference between this splitting and the even one is generally within the experimental error of the determinations of $\Delta_t G^\infty$(i, W→S), and can be ignored unless very accurate determinations of all the involved Gibbs energies are made. In this respect the TPTB assumption is superior to the TATB one, since $V_{vdW}(Ph_4P^+)/V_{vdW}(BPh_4^-) = 1.0218$, i.e., nearer unity, although TPTB has been employed less frequently.

More serious is the objection made on account of the rather open, propeller-shaped, structures of the tetraphenyl ions. Small solvent molecules, such as those of water or formamide, can penetrate between the phenyl rings and interact specifically with the charges. Space-filling models of the tetraphenyl ions show that the approach to the central atom is not very close, however [63]. In addition, the highly polarizable phenyl rings themselves, permitting some delocalization of the central charge, interact with solvents in a slightly different manner when attached to the positively and negatively charged central atoms. This was found by the sensitivity of the NMR signals of ^{1}H and ^{13}C in the *para*-position [74] to the nature of the solvent, suggesting some different hydrogen bonding of the anion with protic solvents. It would require ab initio quantum chemical computer calculations of the energetics of interactions of the constituent cations and anions of TPTB and TATB with solvent molecules to establish how large the qualitatively acknowledged differences between their solvation energies actually are. Such computations appear to be beyond present day capabilities but could be a challenge for the future.

For enthalpies of transfer, measurements of the heats of solutions of electrolytes in W and in S provide the most accurate values, the probable errors being ~±0.3 kJ mol^{-1}. Their splitting into the individual ionic contributions has been made by the TATB assumption, no better method having been suggested [59]. In fact, if the TATB assumption is accepted for $\Delta_t G^\infty$(i, W→S) at 298.15 K, as is widely practiced, there is no reason to confine the acceptance to this one temperature. Therefore, this assumption must also be accepted as a corollary for $\Delta_t H^\infty$(i, W→S) and $\Delta_t S^\infty$(i, W→S). However, the "if" in the former sentence still stands.

In retrospect, the objections to the reference electrolyte extra-thermodynamic assumption for obtaining individual ionic values as implemented by the TATB and TPTB reference electrolytes are serious, but not fatal. Even splitting may introduce tolerable errors (i.e., within the experimental errors) for such quantities where the values for the constituent ions are small, such as enthalpies and Gibbs energies of hydration and equivalent conductivities [31, 63]. The taking into account, where possible, of the known differences in sizes of the cation and anion to produce an asymmetrical splitting is seen to permit the use of this assumption for the molar volumes. However, in cases where the property of TATB or TPTB, such as the molar heat capacity or the viscosity *B*-coefficient, is much larger than for electrolytes made up from small ions and where there is no way to take into account the discrepancy in size, such reference electrolytes should be shunned, provided that better assumptions are available.

CONCLUSION

One essential development desirable in the future, relevant to the properties of individual ions, is a theory of ion-ion interactions applicable in non-dilute solutions. Such a theory should permit the *prediction*, rather than the *modeling*, of activity coefficients, densities, and related quantities of electrolytes at concentrations from 1 mol dm^{-3} upwards.

Another desirable development is the obtaining of standard molar individual ionic thermodynamic quantities at infinite dilution in any one solvent with extrathermo-dynamic assumptions less prone to objections than the present arsenal provides. Water is the obvious choice for this one solvent, due to its importance and the already available information, the well-established conventional values. The relevant quantities for other solvents would then depend on proper assumptions for the transfer between this reference solvent and the target one.

Concerning the transfer functions, the quantum chemical computations of the different interactions of the cation and anion of a reference electrolyte with solvents is a challenge for the future. Another challenge lies in finding a better reference electrolyte, where the differences are even smaller than those thought to exist for the ions of TATB or TPTB. Branched aliphatic chains rather than phenyl rings may better prevent small solvent molecules from approaching and specifically interacting with the buried charges. Such chains would also be less polarizable and have negligible partial charges in the periphery, to interact with solvent molecules. Fluorine rather than hydrogen atoms at the periphery of the reference ions would increase the inertness. Whether such reference ions differing only in the signs of the central, well sequestered charges, can be synthesized and be stable in solutions of interest remains an open question.

REFERENCES

[1] Debye, P.; Hückel, E. *Phys. Z.* 1923, *24*, 185-206, 305-325.

[2] Harned, H. S.; Owen, B. B. *The Physical Chemistry of Electrolytic Solutions*, Reinhold: New York, NY, 1958, 3rd ed.; Robinson, R. A.; Stokes, R. H. *Electrolyte Solutions*, Butterworths, London, 1965, revised 2nd ed.

[3] Kielland, J. *J. Am. Chem. Soc.* 1937, *59*, 1675-8.

[4] Pitzer, K. S. in *Activity Coefficients in Electrolyte Solutions*, Pytkowicz, R. M., Ed., CRC Press: Boca Raton, FL, 2nd edn., 1991.

[5] Marcus, Y.; *J. Solution Chem.*, 2006, *35*, 1271-1286.

[6] Conway, B. E. *J. Solution Chem.* 1978, *7*, 721-770.

[7] Guggenheim, E. A. *J. Phys. Chem.* 1929, *33*, 842-849.

[8] (a) Weingartner, H.; Schroer, W. *Adv. . Chem. Phys.* 2001, *116*, 1-66. (b) D'Angelo, P.; Pavel, N. V. *J. Synchrotron Rad.* 2001, *8*, 173-177.
[9] Rode, B. M.; Hofer, T. S. *Pure Appl. Chem.* 2006, *78*, 525-539.
[10] Marcus, Y. *Chem. Rev.* 1988, *88*, 1475-1498.
[11] Soper, A. K.; Neilson, G. W.; Enderby, J. E.; Howe, R. A. *J. Phys. C Solid St. Phys.*, 1977, *10*, 1793-1801.
[12] David, F.; Vokhmin, V.; Ionova, G. *J. Mol. Liquids* 2001, *90*, 45-62.
[13] Penner-Hahn, J. E. *Coord. Chem. Rev* 1999, *190-192*, 1101-1123.
[14] Barthel, J. *J. Mol. Liquids* 1995, *65/66,* 177-85.
[15] Ishiguro, Sh-i; Umebayashi, Y; Kanzaki, Anal. Sci. (2004), *20*, 415-421.
[16] (a) Sigel, H. *Applications of NMR to Paramagnetic Species* Sigel, H., Ed.; Metal Ions in Biological Species, 1987, Vol. 21, 1-295. (b) Delpuech, J. J. *Analyst* 1992, *117*, 267-88.
[17] (a) Nernst, W. *Z. Phys. Chem.* 1888, *2*, 613-621. (b) Hartley, G. S *Phil. Mag.* 1931, *12*, 473-88.
[18] (a) Fuoss, R.; Hsia, K.-L. *Proc. Natl. Acad. Sci. U.S.* 1967, *57,* 1550-1557 ; (b) Fernandez-Prini, R. *Trans. Faraday Soc.*1969, *65*, 3311-3313.
[19] Marcus, Y. *Elektrokhimiya (English Edition)*, 2007 in the press.
[20] Zana, R.: Yeager, E. *J. Phys. Chem.* 1967, *71*, 521-536.
[21] Marcus, Y.; Hefter, G. *Chem. Rev.*, 2004, *104*, 3405-3452.
[22] (a) Randles, J. E. B. *Trans. Faraday Soc.* 1956, *52*, 1573-1581 ; (b) Case, B.; Parsons, R. *Trans. Faraday Soc.* 1967, *63*, 1224-1239.
[23] Trasatti, S. *J. Chem. Soc., Farady Trans. 1* 1974, *70*, 1752-1768.
[24] Parfenyuk, V. I. *Koll. Zh.* 2002, *64*, 651-659; *Coll. J.* 2002, *64*, 588-595.
[25] Tuttle, T. R. Jr.; Malaxos, S.; Coe, J. V. *J. Phys. Chem. B* 2002, *106*, 925-932.
[26] (a) Kelly, C. P.; Cramer, C. J.; Truhlar, D. G. . *J. Phys. Chem. B* 2006, *110*, 16066-16081; (b) Kelly, C. P.; Cramer, C. J.; Truhlar, D. G. . *J. Phys. Chem. B* 2007, *111*, 408-411.
[27] Marcus, Y. *Biophys. Chem.* 1994, *51*, 111-127.
[28] Halliwell, H. F.; Nyburg, S. C. *Trans. Faraday Soc.* 1963, *59*, 1126-40.
[29] Marcus, Y. *Ion Properties*, Marcel Dekker: New York, NY, 1997.
[30] Marcus, Y. *J. Chem. Soc., Farady Trans. 1* 1987, *83*, 339-349.
[31] Marcus, Y. *J. Chem. Soc., Farady Trans. 1* 1987, *83*, 2985-2992..
[32] Schmid, R.; Miah, A. M.; Sapunov, V. N. *Phys. Chem. Chem. Phys.* 2000, *2*, 97-102.
[33] Cox, B. G.; Parker, A. J. *J. Am. Chem. Soc.* 1973, *95*, 6879-6884.
[34] Marcus, Y.; Hefter, G.; Chen, T. *J. Chem. Thermodyn.* 2000, *32*, 639-649; Lebedev, B. V.; Smirnova, N. N.; Tsvetkova, L. Ya.; Marcus, Y.; Hefter, G. T. *J. Chem. Thermodyn.* 2001, *33*, 485-498.
[35] Millero F. J. *Chem. Rev.* 1971, 71, 147-176.
[36] Hefter, G.; Marcus, Y. *J. Solution Chem.* 1997, *26*, 249-266.
[37] Hedwig, G. R.; Hakin, A. W., *Phys. Chem. Chem. Phys.* 2004, *6*, 4690-4700.
[38] Conway, B. E.; Verrall, R. E.; Desnoyers, J. E. *Trans. Faraday Soc.*, 1966, 62, 2738-2749
[39] Verrall, R. E.; Conway, B. E. *J. Phys. Chem.* 1966, *70* , 3961-3969.

[40] Jolicoeur, C.; Philip, P. R.; Perron, G.; Desnoyers, J. E. *Can. J. Chem.*, 1972, *50*, 3167-3178.

[41] Hefter, G. T.; Marcus, Y. *Chem. Rev.* 2004, *104*, 3405-3452.

[42] Abraham, M. H.; Marcus, Y. *J. Chem. Soc., Faraday Trans. 1* 1986, *82*, 3255-3274.

[43] Jenkins, H. B. D.; Marcus, Y. *Chem. Rev.* 1995, *95*, 2695-2724.

[44] Gurney, R. W. *Ionic Processes in Solution*, McGraw-Hill, New York, NY, 1953.

[45] Parker, A. J. *Chem. Rev.* 1969, *69*, 1-32.

[46] Cox. B. G. *Ann. Rep. Chem. Soc. (London)* 1973, *70*, 249-274.

[47] Marcus, Y. *Pure Appl. Chem.* 1983, *55*, 977-1022.

[48] Persson, I. *Pure Appl. Chem.* 1986, *58*, 1153-1161.

[49] Gritzner, G. *Pure Appl. Chem.* 1990, *62*, 1839-1858.

[50] Marcus, Y. *Pure Appl. Chem.* 1983, *55*, 977-1022.

[51] Waghorne, W. E. *Chem Soc. Rev.* 1993, *22*, 285-292.

[52] Marcus, Y. *Pure Appl. Chem.* 1990, *62*, 799-840.

[53] Marcus, Y. *Electrochim. Acta*, 1998, *44*, 91-98.

[54] Kalidas, C; Hefter, G.; Marcus, Y. *Chem. Rev.* 2000, *100*, 819-852.

[55] Y. Marcus *Z. Naturf. A*, 1995, *50A*, 51-58.

[56] Y. Marcus *Chem. Rev.*, 2007, *107*, in press.

[57] Marcus, Y. *Pure Appl. Chem.* 1985, *57*, 1103-1129

[58] Gritzner, G. *Electrochimica Acta* 1998, *44*, 73-83.

[59] Hefter, G. T.; Marcus, Y.; Waghorne, W.E. *Chem. Rev.* 2002, *102*, 2773-2836.

[60] Marcus, Y. *Pure Appl. Chem.* 1996, *68*, 1495-1500.

[61] Marcus, Y.; Hefter, G. T. *J. Chem. Soc., Faraday Trans.* 1996, *92* , 757-761.

[62] Marcus, Y.; Hefter, G.; Pang, T.-S. *J. Chem. Soc., Faraday Trans.*, 1994, *90*, 1899-903.

[63] Marcus, Y. *Pure Appl. Chem.*, 1986, *58*, 1721-1736.

[64] Parker, A. J.; Alexander, R. *J. Am. Chem. Soc.* 1968, *90*, 3313-3319.

[65] Cox, B. G.; Parker, A. J.; Waghorne, W. E. *J. Am. Chem. Soc.* 1973, *95*, 1010-1014.

[66] Villermaux, S.; Delpuech, J. J. *J. Chem. Soc., Chem. Comm.* 1975, 478-479.

[67] Marcus, Y. *Rev. Anal. Chem.* 2004, *23*, 269-302.

[68] Koepp, H. M.; Wendt, H.; Strehlow, H. *Z. Elektrochem.* 1960, *64*, 483-491.

[69] Gritzner, G. *Inorg. Chim. Acta* 1977, *24*, 5-12.

[70] (a) Krishtalik, L. I. *Elektrokhimiya* 2006, *42*, 1118-1130; *Russ. J. Electrochem.*2006, *42*, 1006-1016; (b) Khanova, L. A.; Topolev, V. V.; Krishtalik, L. I. *Chem. Phys.* 2006, *326*, 33-42.

[71] Grunwald, E.; Baughman, G.; Kohnstam, G. *J. Am. Chem. Soc.* 1960, *82*, 5801-5811.

[72] Alexander, R.; Parker, A. J. *J. Am. Chem. Soc.* 1967, *89*, 5549-5551.

[73] (a) Kim, J.-I. *J. Phys. Chem.* 1978, *82*, 191-199. (b) Kim, J.-I. *Z. Phys. Chem. (NF)* 1978, *113*, 129-150.

[74] (a) Coetzee, J. F.; Sharpe, W. R. *J. Phys. Chem.* 1971, *75*, 3141-3146; (b) Farmer, R. M.; Sasaki, Y.; Popov, A.I. *Austr.J. Chem.* 1983, *36*, 1785-1792.

In: Solution Chemistry Research Progress
Editor: Darian V. Bostrelli, pp. 69-92
ISBN 978-1-60021-993-1

Chapter 3

MOLECULAR RECOGNITION OF SUGARS BY MACROCYCLIC RECEPTORS IN SOLUTIONS

Elena V. Parfenyuk*

Institute of Solution Chemistry of Russian Academy of Sciences, 1 Akademicheskaya Str., Ivanovo, 153045, Russia

ABSTRACT

Molecular recognition of sugars serves as a basic concept to understand biological functions of these compounds and offers a key in developing new types of sugar-based functionalized materials. Since most sugars are isomers which only differ in the stereo arrangements of specific OH-groups, in general the complex formation will be determined by molecular recognition of their structure but not distinctions of chemical reactivity. From this point of view, sterically well-organized macrocyclic compounds such as crown ethers, cyclophanes, cyclodextrines, *etc.* are the most attractive as potential receptors for sugars. This review includes the results reported in literature and own data concerning an ability of mono- and oligosaccharides to form noncovalent intermolecular complexes with natural and synthetic macrocyclic compounds in solutions. Much attention is paid to the effect of stereochemical structure of the sugars on the selective interactions with the macrocyclic ligands. It is an attempt to elucidate the nature of the interactions responsible for molecular recognition of sugars and the role of solvent in this process.

* Fax: +7 4932 336237; E-mail: evp@isc-ras.ru

INTRODUCTION

Molecular recognition of sugars plays an essential role in different important biological events occurring in living organisms such as immune response [1,2], viral and bacterial infections [3,4], drug activity [5], metastasis of tumor cells [6,7], *etc.* In the present time a concept of the sugar code is proposed [8,9]. According to the concept, monosaccharides are a third biochemical alphabet (after nucleic acids and proteins) for encoding biological information. Such information is read by the appropriate receptors to trigger the intended biological response. Therefore, selective interactions of sugars with compounds of a different nature and structure have been the focus of increased study during the last decades. This concerns macrocyclic compounds too.

The purposeful studies on complex formations between sugars and macrocyclic ligands are done for by several reasons. First of all, the macrocyclic compounds provide successful models mimicing various biochemical processes [10,11]. Another reason is to search for new synthetic receptors for the purpose of creating sugar sensors [for instance, 12-14]. The macrocyclic compounds can play a role in both the recognition and signaling parts of the sensors. Besides, owing to their ability to selective interactions with sugars, these compounds are applied for the separation of various sugars and their isomers [15,16].

Since most sugars are isomers that only differ in the stereo arrangements of specific OH-groups, in general the complex formation will be determined by molecular recognition of their structure but not distinctions of chemical reactivity. Therefore, besides required binding sites, potential receptors should be sterically well-organized. Since complexes between large organic molecules form due to noncovalent interactions (van der Waals, electrostatic, dipole-dipole interactions, hydrogen bonding, hydrophobic effects), an availability of multiple binding sites is required to provide high structural organization of the complexes formed. Energy of one noncovalent binding is too small in comparison with that for a covalent bond. From this point of view, the macrocyclic compounds such as cyclodextrins, cyclophane, crown ethers, *etc.* are the most attractive. An availability of a cavity of certain size as well as multiple binding sites provides subtle stereospecific recognition of sugar molecules.

In the present work, an attempt of generalization of literature and my own studies on intermolecular complex formation between mono- and oligosaccharides as well as their derivatives and macrocyclic ligand in solutions for the last fifteen years has been made. The aim of this work is to understand the structural origin of the affinity and selectivity of the interactions and the contribution of the different polar and nonpolar interactions into the stability of the complexes as to well as elucidate the role of a solvent in these processes.

MOLECULAR COMPLEXES OF SUGARS AND THEIR DERIVATIVES WITH MACROCYCLIC COMPOUNDS IN SOLUTIONS

1. Complexes with Nitrogen Containing Macrocycles and their Metallocomplexes

Porphyrins and their analogs are widely extended in nature and the most important representatives among nitrogen containing macrocycles. Porphyrins are various aromatic tetrapyrrole compounds containing a multicircuit conjugate π-system. A sixteen-member macrocycle composed of N and C atoms forms the basis of this system. Porphyrin molecules are planar and have the coordination cavity (2 Å) formed by nitrogen atoms. They have low conformational flexibility. The reaction sites of porphyrin molecules are shielded by π-electron cloud (strict macrocyclic effect) [17]. This effect prevents interactions of the macrocycle with other particles including an intermolecular complex formation with sugars.

However some porphyrin molecules have functional peripheral substitutes which may serve as binding sites for sugars. Therefore a lot of studies have been devoted to a search for an optimal design of the porphyrin receptors to the molecular recognition of sugars. A main direction of such search is the modification of the macrocycles by "hanging" of different substitutes [18-28] or by conjugating with other molecules [29,30] that enable selective binding of sugars. Phosphonate groups, boronic acid groups as well as sulfo-groups are known to be a strong hydrogen bond acceptor and can efficiently bind sugars in water and water-organic solvents [18-25]. Sugars can affect aggregation of the porphyrins containing the above- mentioned polar groups [24-26]. Porphyrins are known to form aggregates in solutions in particular due to cationic –anionic or π-π interactions [31,32]. The complex formation with sugars changes the aggregation state of the porphyrins and their optical properties. Introduction of voluminous peripheral aromatic substitutes into the macrocycles promotes the complex formation with sugars due to noncovalent interactions (hydrogen bonding, hydrophobic interactions) [13, 23, 33]. The chromophore system in these receptors plays the role of signaling group indicating the complex formation with sugar.

However in the cases mentioned above, the complex formation occurs without direct participation of the macrocycles themselves. From this point of view, metallated porphyrins are the most interesting. The metal complexes of porphyrins (MP) are capable of extracoordination, that is additional binding of ligands by central metal ion. This results in coordination saturation of the chemical affinity of the ion. The phenomenon of extracoordination plays an important role in the complexes of natural porphyrin ligands in relation to their biological functions (chlorophyll, B_{12} vitamin, blood heme, cytochromoxidases, *etc.*). The central metal ion (for example, zinc) has an excess positive charge and severs one of a number of binding sites for sugar. This is confirmed by red shift of the Soret peak maxima and the difference spectra showed isosbestic points of metalloporphyrins in chloloroform [34]. Besides, when the zinc of metalloporphyrin *1* is replaced with two protons, the binding constant of β octyl glucopyranoside with the free base of *1* is much reduced (K_{MP} = 41400 M^{-1}, K_P = 600 M^{-1}). This confirms the important role of the zinc in the complex formation. Similarly, the binding constants between Zn(II)porphyrin conjugated with steroid fragment and some pyranosides in apolar solvent [35] show the importance of the central metal ion for sugar binding although free base of the metalloporphyrin binds the pyranosides almost as effectively as Zn(II) porphyrin, but in different orientation.

But studies on interactions between urea-functionalized porphyrins with amino acid substituents and sugar derivatives indicate that the zinc and free base porphyrins have similar binding constants suggesting that the role of central metal ion as binding site for sugars is not essential [28].

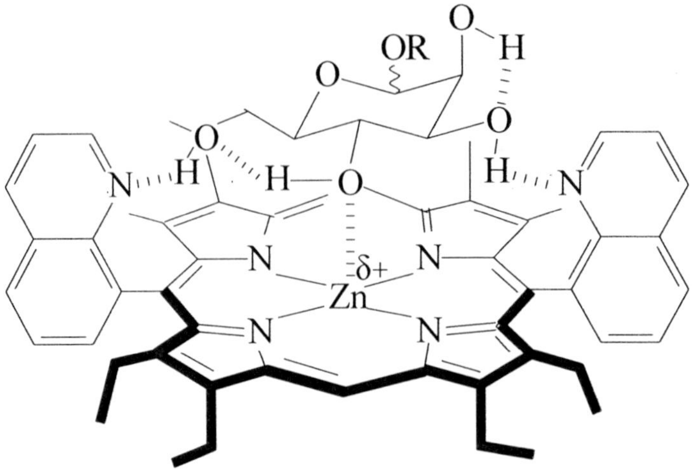

1

However the central metal ion of metalloporphyrin is not only interaction site for sugars. To provide a high structural organization of complexes forming between large organic molecules, multiple binding centers are necessary. Therefore, NMR studies showed fixation of pyranoside molecule above porphyrin receptor plane due to both donor/acceptor interaction with metal ion and noncovalent interactions with peripheral substitutes of the metalloporphyrin [34, 35]. Such complex is formed in the case of the metalloporphyrin containing quinolyl groups fixed at 5-and 15-positions of porphyrin with a *cis* configuration *1*. For the *trans* isomer, the two-point interaction is observed because only one quinolyl nitrogen and zinc can interact with ligand.. This may be due to steric discrepancy between the binding sites of the interacting molecules .

An availability of a few binding sites in macrocycle demands a suitable structural organization of sugar molecules. The interacting particles must be complementary to each other according to form, size and positions of binding sites. Watson called this by the principle of selective "adhesiveness" of molecules [36]. Observance of complementary is particular important at the complex formation between sugars and macrocycles because in the most cases hydrogen bonds are major determinants of specificity and affinity in such complexes. As it is known, hydrogen bonds are highly directional. This fact as well as mentioned above multiple binding impose particular geometric requirements on the intermolecular interactions between such large amphiphilic organic molecules as sugars and macrocyclic compounds. Studies on the molecular recognition of sugars by different macrocycles in solutions show that stereochemical structure of sugar molecules has direct influence on the sugar – macrocycle interactions. The results reported in [34, 35] indicate that the stability of complexes of metalloporhpyrin receptors with pyranosides decreases in the order: mannopyranoside > glucopyranoside >galactopyranoside. It should be noted that β-anomers are bound more strongly then the corresponding α-anomers. A crucial role of OH-group at C4 of the pyranosides was revealed at their complex formation with Zn(II)porphyrins *1* [34].It was established that the OH-group at C4 (equatorial position) is directly involved in the Zn-pyranoside interaction and provides the high affinity of mannopyranoside and glucopyranoside to the porphyrin receptors. Galactopyranoside having the axial OH-group at C4 forms less stable complex. However according to the results of study [28], the selectivity of interaction between the pyranosides and zinc urea functionalized porphyrins with amino acid substituents is the reverse of the mentioned above but anomeric selectivity in general is the same with the β anomer being bound more strongly than the α anomer.

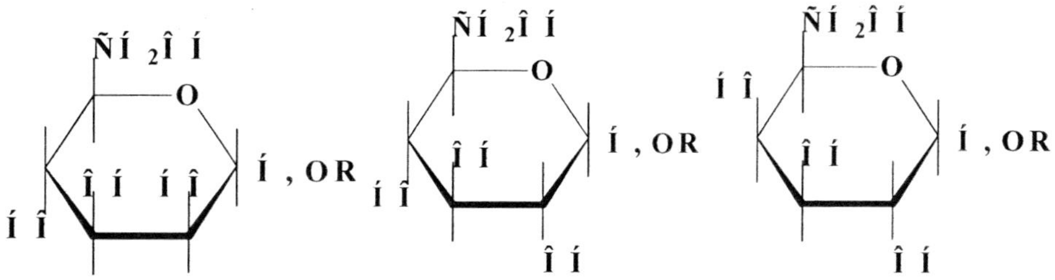

R = $(CH_2)_7CH_3$

D-mannopyranoside D-glucopyranoside D-galactopyranoside

Sugar molecules are able to intramolecular hydrogen bonding in organic solvents. The ability depends on spatial arrangement of OH groups in their molecules [37, 38]. The intramolecular hydrogen bonding in one's turn has influence on the complex formation process. The less intramolecularly hydrogen bonded a pyranoside is, the more hydrophilic ("sticker") it will be where sticky means the ability to form intermolecular hydrogen bonds with a receptor [35]. Equatorial-axial orientation of the OH group at the second carbon atom (C2) in relation to the anomeric oxygen in the α anomer of gluco- and galactopyranosides has stronger intramolecular hydrogen bonds than equatorial- equatorial one in the β anomer. This explains the anomeric selectivity of the pyranosides relatively the metalloporphyrins. As has been noted in study [35], the "stickiness" of sugars soluble in organic media plays an important role at complex formation with any hydrogen bonding receptors for this type of ligand.

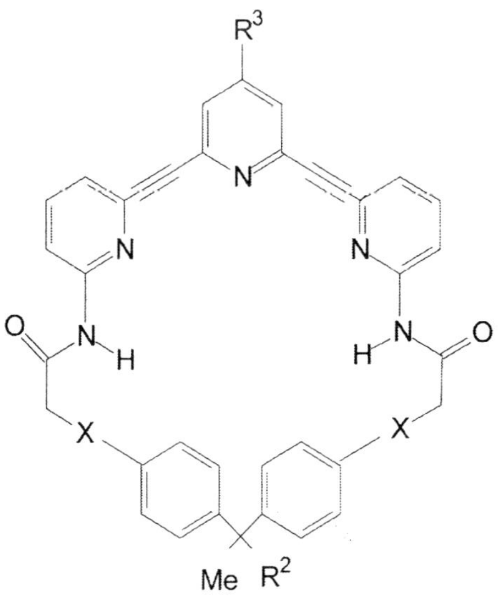

R^2 = Me, R^3 = H, X = O
R^2 = Me, R^3 = H, X = S
R^2 = Me, R^3 = H, X = CH_2
R^2 = Bu-i, R^3 =H, X = CH_2
R^2 = Bu-i, R^3 =OBU-n, X = CH_2

2

Polypyridine-macrocyclic receptors (*2*) are such receptors. The hydrogen bonding between OH groups of the receptors and those of sugars is responsible for the complex formation. Therefore, the selectivity in the binding of sugars by *2* depends strongly on stereochemistry of sugars as well as their shapes [39]. The pyridine nitrogen can serve as a hydrogen acceptor, while the amide NH group as a donor. The receptors have binding cavities of fixed size and flexible structures. Inouye et al [39] showed that the replacement of the oxygen atoms (X) by CH_2 groups leads to the lack of the intramolecular hydrogen bonds X···NH. This makes the receptor molecule more flexible and results in more favorable conformation for the binding of sugar produced a very large enhancement of the binding affinity (50-folded). Only a little or negligible binding affinity of the receptors for hexoses was observed. The most stable complex is formed with methyl β-D-ribofuranoside.

2 Complexes with Cyclophanes

Sugar molecules have a highly polar nature. Therefore the design of new receptor molecules that can precisely recognize sugar molecules employing hydrogen bonding has become a very active area of research. Calixarenes proved to be the most perspective macrocyclic compounds for this purpose. An ability of monosaccharides to complex formation with synthetic polyhydroxy cyclotetramers has been studied by Aoyama et al. [40-42]. The compounds *3a-c* have a symmetric macrocyclic skeleton with intramolecular hydrogen bonding between OH groups of adjacent benzene rings and a bowl-shape cavity. All eight hydroxyl groups are inwardly facing into a cavity. These lipophilic compounds have been shown to extract sugars from water into apolar organic media upon formation of hydrogen-bonded complexes. The formation of stable *3a*-sugar complexes in CCl_4 is attained by two-point interactions of two OH groups of monosaccharide with two pairs of OH groups of adjacent benzene rings bonded by intramolecularly. The selective binding of sugars depends on stereochemistry of monsaccharide molecules. Studying a wide series monosaccharides [40], the authors came to a conclusion that the relative spatial configuration of OH groups at third (C3) and fourth (C4) carbon atoms plays a particular role upon complex formation with *3a-c*. For *cis* stereochemistry of the OH groups of aldoses the affinity of sugar to *3a* is higher than for *trans* configuration. In addition, the spatial position of OH groups as well as nature of substituent at fifth carbon atom (C5) (CH_3 >> H >>CH_2OH) has influence on lipophilicity of the sugar molecule as a whole. The higher lipophilicity , the stronger the complex in apolar solvent. The affinities of to *3a* various aldopentoses, aldohexoses and their deoxy derivatives decrease in the order: fucose (6-deoxygalactose) > 2-deoxyribose > ribose > arabinose ≅ rhamnose (6-deoxymannose)>> galactose ≅ xylose ≅ lyxose ≅ mannose ≅ glucose. The structural requirements for the formation of stable *3a*-sugar complexes in apolar media are explained by maximization of favorable *3a*-sugar interactions and minimization of unfavorable exposure of sugar OH groups to bulk solvent [40,41]. Therefore D-ribose possessing a lipophilic face and hydrophobic face, as result of its D- stereochemical structure, shows a high affinity to *3a* observed both in an apolar solvent and in a monolayer of the macrocycle from dilute subphases at the air-water interface while L-fucose, possessing a methyl group on C5 of the pyranose ring, forms a stable complex with *3a* only in apolar media [41].

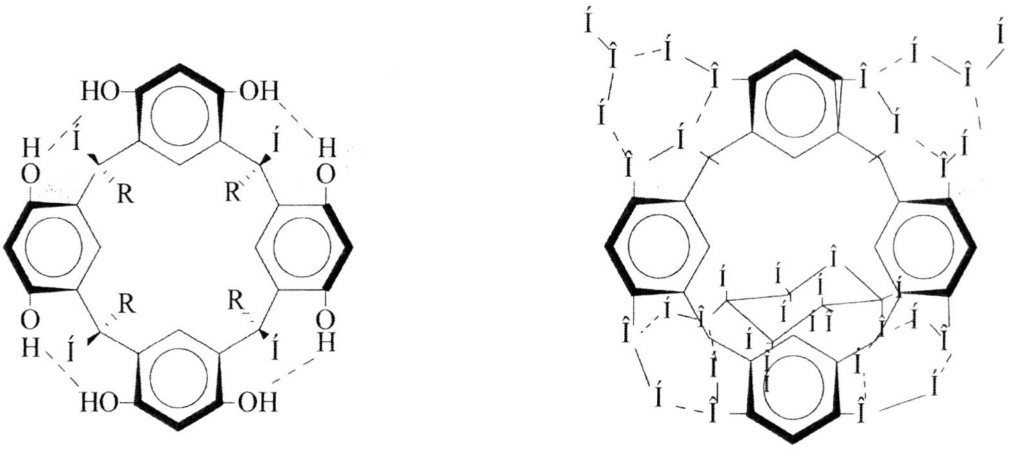

3a-c: 3a R = $(CH_2)_{10}CH_3$; 3b R = $(CH_2)_6CH_3$; 3c R = $(CH_2)_4CH_3$

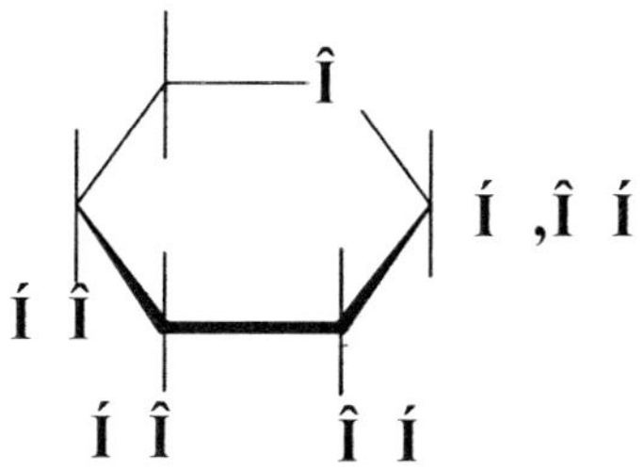

D-ribose

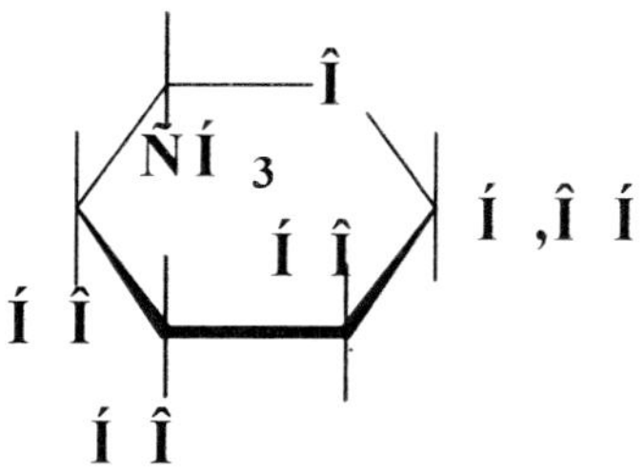

L-fucose

Alkyl substituents in monisaccharide molecules do not change in principle the nature of interactions responsible for complex formation with the calixarene macrocycles. The complex formation of the methyl glucopyranoside and n-octyl glucopyranoside with *3a* in apolar media (CCl_4, $CHCl_3$) is also promoted preferentially by hydrogen bonding interactions [42]. The methyl glucopyranoside is solvophobic or otherwise insoluble in apolar solvents. This glucopyranoside aims to insulate its OH groups from the bulk apolar solvent by undergoing an extensive hydrogen -bonding interaction with the macrocycle. This results in a compact 2 : 1 sugar-encapsulating complex with a remarkable β anomer selectivity. The authors of this work suppose that the selectivity may be connected with equatorial position of the methyl group in the β anomer and hence, a close proximity of this group and a benzene ring of *3a* promoting CH-π interaction. N-octyl glucopyranoside has long alkyl "tail" and is soluble in apolar media. This glucoside forms 1 : 4 complex with the macrocycle (*3a* : sugar): the four sugar molecules are bound at the four unit hydrogen-bonding sites of the macrocycle with a high cooperativity that arises from intracomplex sugar-sugar hydrogen bonding invoving the OH groups at C5 and C2 carbon atoms of the adjacent glucoside molecules. The octyl groups are extended into bulk solvent.

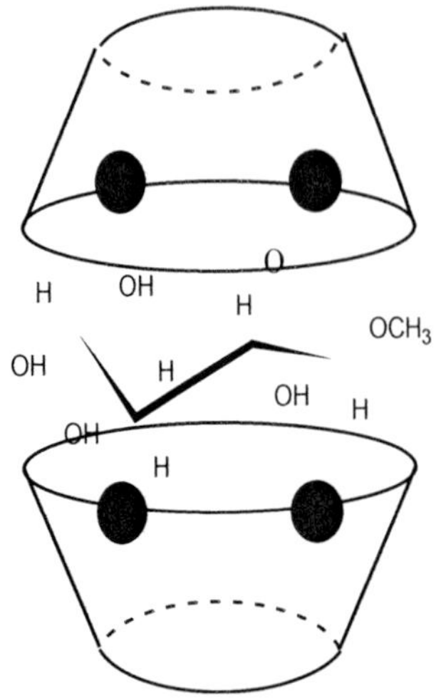

encapsulation complex 2 : 1 (*3a*:sugar)

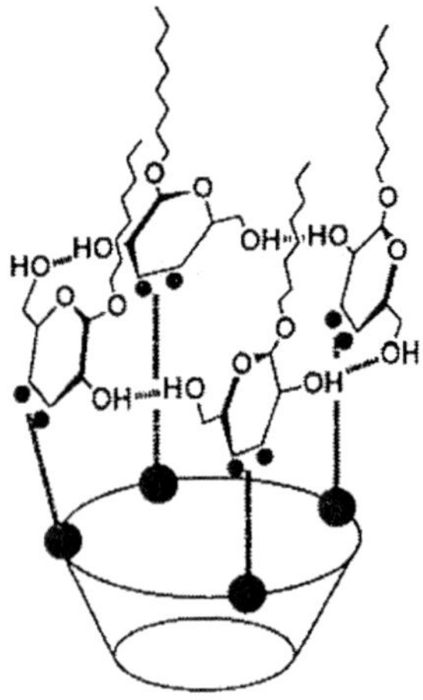

complex 1 : 4 (*3a*:sugar)

Introduction of bulky binaphthyl substituents into calixarene macrocycle *4* play an essential role in the solubility of the macrocycle in organic solvents and the formation of the complexes with sugars [13]. Authors of this work believe that an effective sugar binding by *4* results from the combination of H-donated groups and aromatic rings in the macrocycle molecule. The macrocycles show high affinities to di- and trisaccharides in water or water-organic mixtures. The selectivity to the β-D-lactose was observed. On the basis of the binaphthyl substituted calixarene the new effective optical system for chemical sensing of sugars in aqueous media was developed [43]. The principle idea of this development is based on a competition mechanism of color label (methyl red) and sugar for binding sites of the receptors. This process is accompanied by color transition which allows sugar binding to be monitored visually.

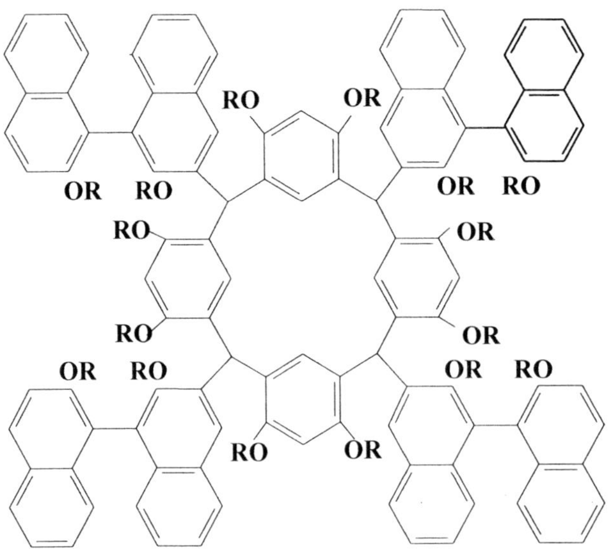

4: R = Me, H

In view of biological molecular recognition of sugars, much attention was paid to the sugar binding receptors working in aqueous media. The water soluble tetrasulfonate derivatives of *3a* [R=$(CH_2)_2SO_3Na$ and different substituents (H, CH_3, OH) at C2 carbon atoms in benzene rings] were synthesized and selective binding of various sugars to these receptors were studied [44]. It was found that unmodified sugars show low affinity to the tetrasulfonate derivatives *3a.* Relatively hydrophobic monosaccharide derivatives (some aldopentoses, and deoxy and methylated sugars) are bound to the receptors although the obtained binding constants are small. Due to their stereochemistry sugars have both hydrophilic and hydrophobic characters [45-47], the latter becoming more important upon deoxygenation and alkylation. However, the authors of this study suppose that the hydrophobic forces or the van der Vaals interactions are not the sole governing factor upon the complex formation .The ^{1}H NMR data have showed that the hydrophobic moiety of sugar molecule is incorporated in the polyhydroxy aromatic cavity of the receptor. It was concluded that CH-π interaction involving electron –rich benzene rings is at least partially responsible for the complex formation.

Cyclophane receptor *5* exists as salts in solution. The solubility of tetracation *5* in water or organic solvents is strongly dependent on the kind of its counter-ions. This receptor has a rigid structure and hydrophobic cavity (10.3×6.8 Å). Originally cyclophane *5* was designed as receptor for binding electron-rich aromatic residues in organic and aqueous solution [48]. An ability of the receptor to bind phenyl glycopyranosides (D-glucose, D-mannose, D-galactose) in aqueous solution was studied [49]. According to the ^{1}H NMR data cyclophane *5* forms 1 : 1 associates with the mentioned sugars. The association is driven by inclusion of the sugar phenyl group inside the cavity of the cyclophane. Sugar anomeric selectivity was observed: the β anomer was found to bind more strongly than the α anomer. Molecular dynamics simulations and Monte Carlo calculations have showed that the anomeric selectivity does not result from differences in electron density on the phenyl rings of the anomers or differential salvation of the two anomers. The preference of the β anomer to the α anomer is steric in nature [50]. An equatorial phenyl group of the β anomer is more well encapsulated by the macrocycle than an axial group of the α anomer. This results in the formation of more stable associate for the β anomer. Although in this case the sugar molecules do not participate directly in binding by cyclophane *5*, these studies indicate clearly the important role of stereochemical structure of sugars in their selective binding by macrocyclic receptors.

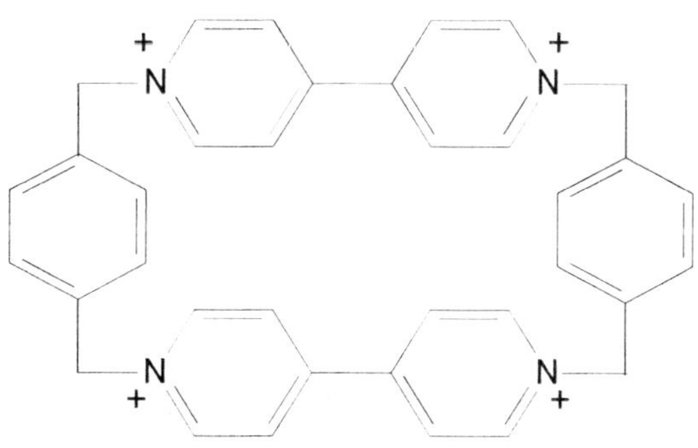

5

3. Complexes with Cyclodextrines

Cyclodextrins (CD) *(6)* are the most "kindred" to sugars in nature and composition. Cyclodextrines are macrocyclic, non-reducing maltooligosaccharides composed of glucose units (not less six) linked by α-(1,4) glycosidic bonds. Each cyclodextrin is a torus (doughnut-shaped) molecule. The internal cavity of the molecule is hydrophobic, whereas the external surface is hydrophilic. This molecule shape is very convenient for modeling of stereospecific enzymatic reactions [11, 51-53], molecular encapsulation of reagents to change their reactionary ability and selectivity [52,53], enantiomer separation [54], *etc.* Therefore, numerous theoretical and experimental works devoted to investigation of interactions of various organic and inorganic molecules with CDs in solution have been reported in literature. Against a background of such great number of the studies, the works concerning binding of sugars by CDs are very rare. This fact may be explained by a large number of works testifying about formation of low-stability complexes between CDs and sugars in aqueous solution [55-57]. The binding constants of sugars to CDs are in the range from unit to several tens ($l \cdot mol^{-1}$). Authors of work [55] believe that this is due to strong hydration of the interacting particles. The complex formation of monosaccharides to CDs (exothermic effect) must be accompanied by strong desolvation of both the CD and the sugar molecules (endothermic effect), leading to a small enthalpy change of this process. The destruction of the ordered water structure in hydration shells and the release of these water molecules to the bulk upon the complex formation result in favorable entropic contribution [55].Thus, the entropic changes from the solvent reorganization give a major contribution to the stability constants of the complexes formed between CDs and monosaccharides.

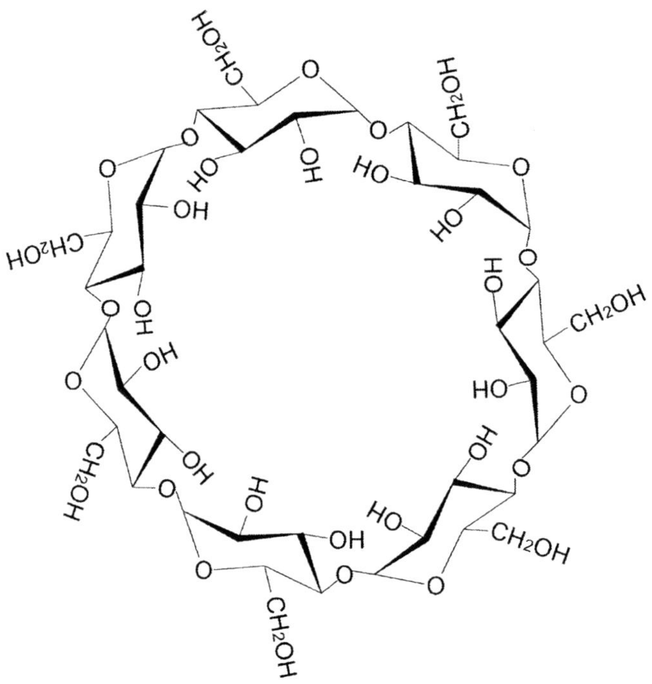

6 (β-cyclodextrin)

The results reported in [55] show that β-CD is able to recognize aldopentoses but no aldohexoses in water. The authors of this study attributed the lack of interaction of this macrocycle with aldohexoses to steric factors, suggesting that the presence (aldohexoses) or the absence (aldopentoses) of a substituent in the C5 of the pyranose ring may play a key role upon the complex formation. In general, α-CD forms more stable complexes with monosaccharides in comparison to β-CD. The ^{13}C NMR studies have shown that the ring oxygen and the axial OH-group at C3 of the monosaccharides are responsible for the sugar-CD interactions. But it should be noted that α-CD is unable to recognize these monosaccharides selectively as assessed from stability constant data.

The preferable binding of sugars to α-CD in comparison to β-CD has been found also in works [58-59] in which these macrocycles have been proposed as bonded phase in chromatographic column s for separation of sugars. The results of these works show that the three parameters, hydration number, number of hydroxyl groups and conformation of sugar molecules do have an effect on the efficiency of their binding to CDs. But it should be noted that aqueous solutions of acetonitrile, methanol and acetone served as mobile phases in the columns.

The complex formation process depends on structure and state in solution of both interacting particles. The nature of heteroatoms in the analogous macrocycles affects directly the molecular complex formation with sugars. For instance, aminocyclodextrins (amino-CD) forms more stable complexes with ribose in comparison to unmodified CDs [56].

The nature of the interactions responsible for the sugar-C D complex formation is still controversial. CDs are known to recognize a wide variety of organic as well as inorganic molecules, forming inclusion complexes in aqueous solution. The hydrophobic cavity of CD on the one hand and hydrophobicity of sugar molecules on the other hand assume that the hydrophobic interaction is a partial driving force for the complex formation. The inclusion complex formation between sucrose and β-CD in aqueous solution is attributed to the cohesive interactions but hydrophilic interactions may also play some part [60]. According to the results obtained in this work, the formation of inclusion complexes between sucrose and β-CD with 1 : 2 ("sandwich") and 1 : 1 stoichiometry have been found. However, another opinion about the nature of sugar-CD interactions gives in the mentioned above studies [56, 58-59]. The most likely explanation of sugar binding to CDs is the formation on intermolecular hydrogen bonds. The protonated amino groups in amino-CD are good donors. Therefore the amino-CD provides much better binding ribose in comparison to the electroneutral CD as well as ribose in comparison to deoxyribose in spite of its higher hydrophobocity, pointing out the hydrogen bonds as the dominant driving force for the sugar complex formation [58]. Armstrong et al. also believe that retention of saccharides on CD bonded phases is due to adsorption via hydrogen bonding but not to inclusion complex formation [58,59].

4. Complexes with Glycophanes

As has been noted above, the combination of proton donated groups and aromatic rings in macrocycle molecule provides an efficient binding of sugars and their derivatives in solution. Search of new artificial receptors working in aqueous media has led to design of glycophanes. Glycophane is cyclodextrin-cyclophane hybrid which is endowed with the properties of both cyclodextrins (chirality, neutrality, hydrophobic cavity and water solubility) and cyclophanes (additional aromatic interactions). Glycophane molecule is constituted by two sugar segments and two aromatic segments (7). The sugar segment is disaccharide such as α,α-trehalose [61-62] or maltose [63]. The aromatic segment can be composed by naphthalene [61] and its derivatives [62] as well as xylene [62] and benzoic acid derivatives [63]. Thus, glycophane 7, endowed with a hydrophobic cavity and electron-rich walls, is an appropriate model to evaluate the influence of aromatic and hydrophobic interactions on the strength and the selectivity of binding of sugars in aqueous medium. The affinities of p-nitrophenyl glycosides to 7 depend on hydrophobicity of the sugars. The number of CH groups in the α-face of the monisaccharides is determined by their stereochemical configuration. The mannose derivative with the smaller number of OH-groups in the interacting α-face is the most hydrophobic and forms the strongest complex with 7 (K = 265 $M^{-1)}$ while the lowest value, K = 130 M^{-1}, was found for the less hydrophobic glucose [62].

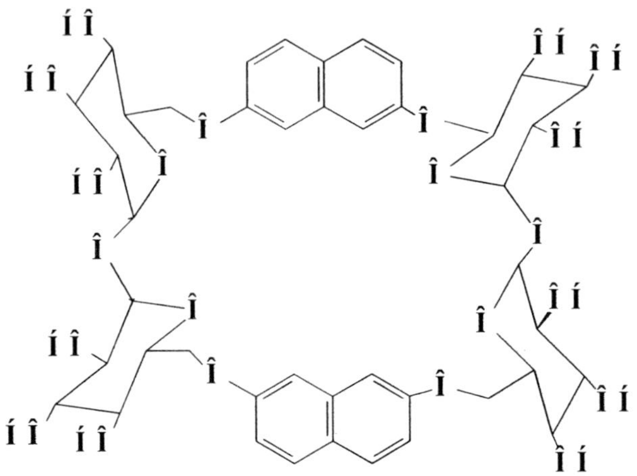

7

Glycophane 7 shows α/β stereoselectivity for these glycosides. The largest selectivity was found for mannose, although in all cases there is preference for the α anomer. The role of the streochemical structure of sugar molecule in the complex formation with glycophanes is emphasized in work [61]. The authors of this work believe that the main driving force for the complex formation is the complementarity of nonpolar surfaces of two interacting particles which is significantly connected with the spatial arrangement of OH groups in sugar molecules. Thus, the role of the hydroxyl groups in a sugar is not only related to the formation of hydrogen bonds with the receptor but also to the strengthening of the hydrophobic interactions. The stability of the complexes mainly is connected with such hydrophobic interactions (van der Vaals and desolvation forces) [61]. A comparison of receptor properties of CD and glycophane 7 in relation to the nitrophenyl derivatives of monosaccharides shows the following distinctions: 1) as has been mentioned above, in contrast to CD, glycophane shows anomeric selectivity; 2) the interaction between sugar moiety of the nitrophenyl glycosides and glycophane 7 contributes to the stability of the complexes formed. The complexes of CD and 4-nitrophenyl glycosides did not show any additional stabilization due to sugar moieties.3) The stereochemistry of the sugar does not seem to influence on the free energy of binding sugar to CD to the same extent than in the case of 7 [61,62] .

5. Complexes with Crown Ethers

In contrast to the examined above cyclodextrins, calixarenes, glycophanes, crown ether molecules (*8*) have hydrophilic cavity and relatively flexible hydrophobic skeleton. Diameter of the cavity depends on size of the macrocycle [64]. The cavity of fixed size and multiple binding sites, as several heteroatoms having electron lone pairs, promote a high complexing ability and selectivity of these macrocyclic ethers with respect to very wide range of organic and inorganic compounds [10, 64-65]. These properties provide an increasing scientific and practical interest to the macrocycles.

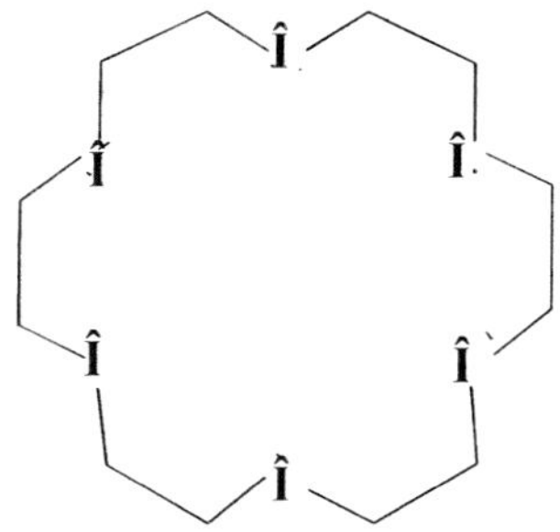

8 (18-crown-6)

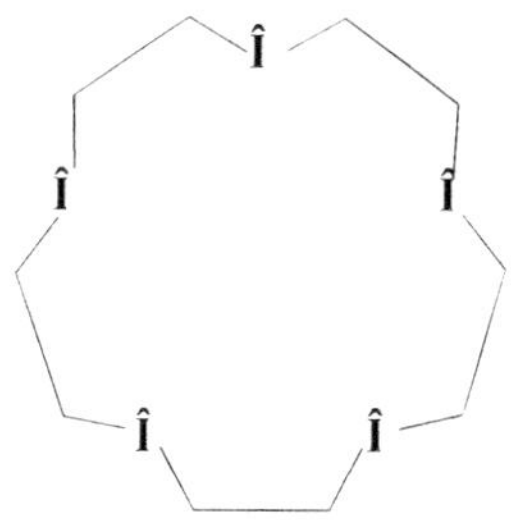

8 (15-crown-5)

The experimental studies of the interactions between mono- and disaccharides and some crown ethers in aqueous solutions [66-76] testify about selectivity of binding of sugars to these macrocyclic compounds (Table 1). Such selectivity is explained by stereochemical structure of the monocassharide molecules. The spatial arrangement of OH groups affects 1) the formation of hydrophobic surfaces of monosaccharide molecule; 2) its hydration; 3) the complementarity of these hydroxyl groups and hydrophilic binding sites of crown ether. One more factor is the size of the macrocycles and their state in solution. The thermodynamic studies [72-74] showed that the first factor has a dominant influence on an ability of two stereoisomeric aldohexoses, D-glucose and D-galactose, to the complex formation with 15-crown-5 and 18-crown-6 in dilute aqueous solution. In contrast to D-glucose, the upper side of D-galactose molecule due to an axial orientation of OH group at C4 is rather polar, the B-face exhibits a hydrophobic character (hydrophobic patches defined by the C3-C6 part).

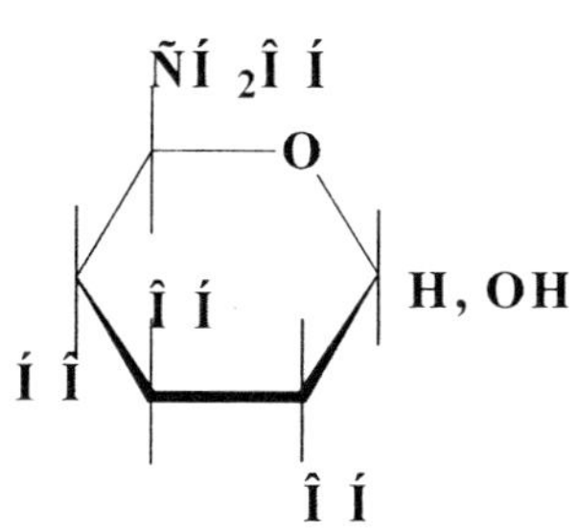

D-glucose

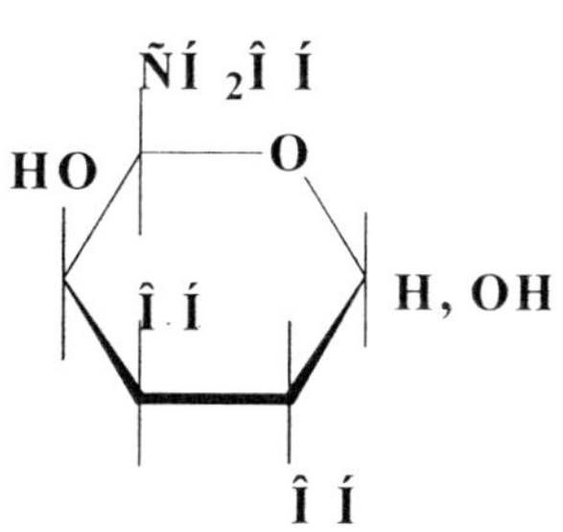

D-galactose

The hydrophobic surface can be a target for van der Waals interactions and hydrophobic effects. Therefore, the complex formation of D-galactose with both the crown ethers is accompanied by heat absorption and contraction. If the configuration of the OH group at C4 is changed, the resulting epimer (D-glucose) will not form the complex neither with 15-crown-5 nor 18-crown-6 due to the distributing introduction of a polar group into hydrophobic environment and /or steric hindrance. Thus, as in the case of the complexes between sugars and glycophanes mentioned above, the spatial orientation of the hydroxyl groups in sugar molecule influences the free energy of the interactions, even though these groups do not seem to be involved in hydrogen bonding with the receptors. The formation of 2 : 1comlex between 15-crown-5 and D-galactose may be connected with the indicated features of stereochemical structure of the monosaccharide molecules as well as state of the crown ether molecule in aqueous solution. According to literature data [77], in aqueous solution the predominant conformation of 18-crown-6 is a centrosymmetric D_{3d} while the asymmetric conformation is characteristic for 15-crown-5 [78]. In addition the crown ether molecules have different cavity size and different hydration energies [79].

An axial orientation of OH group at C4 in L-arabinose molecule as well as the fact that this monosaccharide exists preferentially in α pyranose form in aqueous solution [80], allows to suppose that such stereochemical configuration of the monosaccharide is complementary to the hydrophilic binding sites of 18-crown-6 but not 15-crown-5. The thermodynamic parameters of complex formation and volume changes of this process show that the hydrophilic interactions are dominant upon the complex formation between 18-crown-6 and L-arabinose. Stereoisomer D-xylose does not form the complexes neither with 15-crown-5 nor 18-crown-6 [75]. In general, it may be concluded that the monosaccharides having at least one axial OH group in their molecules form the intermolecular complexes with the crown ethers.

Oligosaccharide molecules are flexible due to rotation of the monosaccharide moieties around glycosidic linkage. Therefore, they can exist in different conformations in aqueous solution including the"folded" conformations formed through intramolecular interactions [81-85]. This leads to an increase of hydrophobicity of oligosaccharide molecules. These factors determine interactions of oligosaccharides with water and macrocycles. According to the data reported in [83,84], α,α-trehalose forms the most compact conformation in comparison to other disaccharides. It is likely that such packed conformation of α,α-trehalose molecule results in more stable complexes between α,α-trehalose and both the crown ethers. It should be noted that these complexes are formed in spite of unfavorable enthalpy changes [70].

Like α,α-trehalose, disaccharides maltose and cellobiose are composed of two glucose moieties in the pyranose form but are linked by different types of glycosidic bonds between the two rings. There are (1-4α) in maltose and (1-4β) glycosidic bonds in cellobiose, respectively. The thermodynamic study [71] showed that both the disaccharides form the complexes with 18-crown-6 but not with 15-crown-5. Maltose and cellobiose form 1 : 2 and 1 : 1 complexes with 18-crown-6, respectively. This may be due to the different types of glycosidic bonds in the disaccharide molecules.

Table 1. Stability constants and thermodynamic parameters of 1 : 1 complex formation between some mono- and disaccharides and crown ethers in aqueous solutions at 298.15 K

sugar	Dominant conformation of monosaccharideor composition of disaccharide	15-crown-5			18-crown-6		
		K, $l \cdot mol^{-1}$	ΔH, $kJ \cdot mol^{-1}$	TΔS $kJ \cdot mol^{-1}$	K, $l \cdot mol^{-1}$	ΔH, $kJ \cdot mol^{-1}$	TΔS $kJ \cdot mol^{-1}$
Monosaccharides							
D-xylose[a]	1e2e3e4e	No complex formation			No complex formation		
L-arabinose[a]	1a2e3e4a	No complex formation			2585±320	-0.33±0.04	19.1
D-ribose	1e2e3a4e	1682±164	-0.27±0.03	17.4	1248±122	-0.22±0.02	18.2
D-glucose[b]	1e2e3e4e6e	No complex formation			No complex formation		
D-galactose[b]	1e2e3e4a6e	Complex 2 : 1 (cr : sug) $2.76 \cdot 10^5 \pm 2.44 \cdot 10^3$	7.80±0.69	38.9	2619±397	3.40±0.52	22.9
Disaccharides							
α,α-trehalose[c]	α-Glc-(1-1)-α Glc	643±44	1.15±0.08	17.3	1295±155	2.08±0.09	19.7
Cellobiose[d]	β-Glc-(1-4)-Glc	No complex formation			$1.45 \cdot 10^4 \pm 1.73 \cdot 10^3$	-1.1±0.1	22.8
β-lactose[c]	β -Gal-(1-4)-Glc	214±7	-1.00±0.16	12.2	No complex formation		
sucrose [c]	α-Glc-(1-2)-β Fru	364±12	-4.16±0.06	10.5	393±15	-4.85±0.05	9.5
maltose[d]	α-Glc-(1-4)-Glc	No complex formation			Complex 2 : 1 (cr : sug) $6.57 \cdot 10^6 \pm 7.13 \cdot 10^5$	-0.8±0.1	38.2

[a] [75].
[b] [73, 74].
[c] [70].
[d] [71].

The replacement of an equatorial OH group at C4 in the first monosaccharide moiety in cellobiose molecule by an axial one leads to significant difference in the behavior of cellobiose and the resulting stereoisomer lactose with respect to the crown ethers (Table 1). These data is one more striking example of delicate stereospecificity of sugars which is a base of important biochemical processes.

The thermodynamic parameters of binding of sucrose with 15-crown-5 and 18-crown-6 do not differ significantly. The complex formation processes between them are accompanied by heat evolution [70]. The studies of volumetric properties of three component system sucrose -18-crown-6-water [86] showed that inclusion of sucrose in the crown ether cavity is likely not possible due to the large size of sucrose molecule in comparison to the cavity. It was assumed that sucrose interacts with 18-crown-6 due to hydrogen bonding between OH groups of the disaccharide and lone electron pairs of oxygen atoms of the crown cavity.

The authors of works [66-76] emphasize a significant role of water upon the complex formation between sugars and crown ethers. The entropic contribution is the major factor governing the complex formation. This is connected with dehydration of the interacting particles and reorganization of the solvent structure upon the complex formation. At the interaction some water molecules from the hydration shells release to the bulk, whereas other water molecules, previously in the bulk phase, may be taken up by the hydration shell of the complex formed. Because the water in the hydration shells is thermodynamically distinct from the bulk water, such reorganization should have a significant influence on the energy of the intermolecular interaction.

6. Complexes with Cholaphanes

Cholic acid is a natural steroid product, an ideal building block for synthetic amphiphilic receptorsm since it has a rigid hydrocarbon skeleton with slight curvature and has hydroxyl groups (three) oriented perpendicular to the steroid plane. If two cholic acid templates ('arms') are connected through terminal OH groups (C3) fitting into vacant space an appropriate geometric fragment (usually, aromatic amines serve as spacer), the forming macrocycle has hydrophilic cavity and hydrophobic outer surface (*9*). Such peculiarities make related cholophane and crown ether molecules. Davis and co-workers [87], synthesized several cholophanes created from cholic acid and benzylamine fragments and having different substituents on the C7 and C12 α hydroxyl groups. The cholic acid units in them are joined head- to-tail. The ^{1}H NMR titration results have showed that these macrocycles are efficient receptors for alkyl glucopyranosides . A glucoside molecule [87] forms six hydrogen bonds with the receptor. Two of these hydrogen bonds are formed between the NH groups of the aromatic spancer and hydroxyl groups of the sugar. New cholophane receptors having different conformations and spacers were designed using molecular modeling method [88]. These cholophanes were designed specially to recognize sugars taking into account stereochemical structure of the latter. The results of the modeling predicted high steric and hydrogen-bonding complementarity of the receptors with sugars [88].

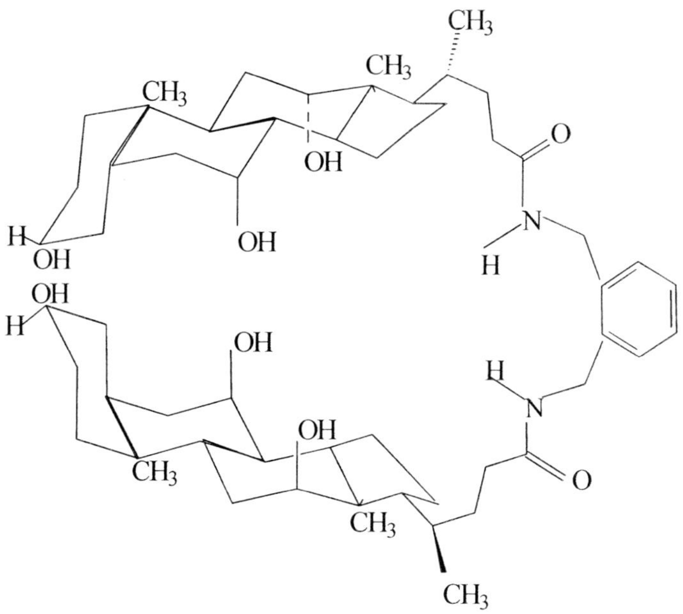

9 (mCAH)

7. Complexes with Cyclic Peptides

Pharmacological effect of some antibiotics is associated with molecular recognition of sugars by these drugs. Well known gramicidin antibiotics possessing antibacterial activity are cyclic peptides. Gramicidin S is a cyclic peptide consisting of 10 amino acid residues serves as a basis to synthesize new cyclic peptides which are able to bind sugars selectively. To strength the hydrogen-bonding between the peptide and sugar residues of the charged amino acids (aspartic acid, asparagine and glutamine) were substituted for valine residues in gramicidin S. The hydrophobic pocket to exert the effect of hydrogen-bonding was prepared by substituting tryptophan redidue for leucine residue [89]. Thus, the following kinds of gramicidin analogues were synthesized:

cyclo(Pro-Aasn-Orn-Trp-D-Phe)$_2$ (*10a*);
cyclo(Pro-Asn-Orn-Trp-D-Phe-Pro-Asp-Orn-Trp-D-Phe) (*10b*);
cyclo(Pro-Asp-Orn-Trp-D-Phe)$_2$ (*10c*);
cyclo(Pro-Gln-Orn-Trp-D-Phe)$_2$ (*10d*).

The data of fluorescence spectroscopy and affinity column chromatography indicated that the peptides possess the selective binding ability for some sugars in aqueous medium. The macrocycle *10a* recognizes monosaccharide fructose but not disaccharide lactulose (Gal- (1-4β) Fru). The replacement of aspartic acid residue by glutamine residue results in an opposite behavior: the macrocycle *10d* prefers the structure of lactulose to that of fructose. In general, the cyclic peptides bind preferentially disaccharide lactulose than fructose. The authors of work [89] could not explain the selectivity against sugars but they noted an important contribution of hydrophobic interactions between the sugars and phenylalanine and tryptophan residues to affinity of the sugars to the macrocycles.

CONCLUSION

Molecular recognition of sugars plays a key role in cell functioning and its "communication" with the external environment. As parts of glycoproteins and glycolipids, cell surface sugars are recognized by different receptors (hormones, toxins, enzymes, drugs, etc.). Their recognition triggers post-binding signaling and the intended biological response. Studies of nature and mechanisms of the interactions responsible for molecular recognition of sugars, establishment of their regularities are of a great scientific interest and applications in pharmacology, immunology, and chemical technology. This review showed that in the present time chemistry of artificial macrocyclic receptors for such biologically important compounds as sugars develops successfully. A wide range of macrocyclic compounds binding sugars selectively both in organic solvents and aqueous media has been proposed. The data reported in the review showed unambiguously that the basis of such delicate specificity of sugars is their stereochemical structure which has influence not only on the spatial complementarity of binding sites of sugar and sterically well-organized macrocyclic receptor but on the formation of apolar surfaces in sugar molecule to optimize the sugar – receptor interactions. Certainly molecular recognition is affected by nature and the structure of macrocyclic receptor. Is it planar or does it have a deep cavity of fixed size. Is it rigid or flexible. However, as has been shown in the review, the most efficient receptors for sugars are the macrocycles composed of both binding sites for specific interactions and apolar fragments which serve as a target for van der Waals interactions. In most cases the selective sugar binding is not a result from one type of interaction. It is characterized, as it is the case with other biological molecules, by the confluence of many weak specific and universal intermolecular forces. Although sugars have a strongly hydrophilic character, the stereochemical structure of monosaccharide molecules as well as the conformations of oligosaccharide molecules in solution determines their lipophilic properties and ability to apolar interaction with macrocyclic ligands. In spite of pronounced growth in a number of studies in this field, it is still difficult to discover regularities which allow predicting the selectivity of sugar – macrocycle interactions. It should be noted that information content concerning this problem is limited in world literature. Summary reviews are rare. Possibly the present review fills a gap in our knowledge in this field at least partially.

REFERENCES

[1] Lehninger A.L. *Principles of Biochemistry.* Worth Publishers, Inc., 1982.

[2] Engström H.A., Andersson P.O. and Olsen S. Analysis of the specificity and thermodynamics of the interaction between low affinity antibodies and carbohydrate antigens using fluorescence spectroscopy. *J. Immun. Methods,* 2005, v. 297, p. 203-211.

[3] Olofsson S., Bergstrom T. Glycoconjugate glycans as viral receptors. *Ann. Med.*, 2005, v.37, N 3, p.154-172.

[4] Williams S.J. and Davies G.J. Protein-carbohydrate interactions: learning lessons from nature. *Trend. Biol.*,2004, v. 19, N 9, p. 356-362.

[5] Okamoto, T.Oki, Y.Igarashi, M.Tsurudome, M.Nishio, M.Kawano, H.Komada, M.Ito, Y.Sakakura, Y.Ito. Enhancement of human parainfluenza virus-induced cell fusion by pradimicin, a low molecular weight mannose-binding antibiotic. *Med. Microbiol. Immunol.*, 1997, v. 186, p. 101-108.

[6] .Gorelik E,.Galili U. and Raz A. On role of cell surface carbohydrates and their binding proteins (lectins) in tumor metastasis. *Cancer Metast. Rev.*, 2001, v. 20, p.245-277.

[7] Dennis J.W. N-linked oligosaccharides processing and tumor progression. In:Fukuda M. (ed.) *Cell Surface Carbohydrates and Cell Development,* CRC Press, Boca Raton, 1992, p.161-194.

[8] Gabius J.-H. Biological information transfer beyond the genetic code: The Sugar code. *Naturwissench.*, 2000, v. 87, p. 108-121.

[9] Sharon N., Lis H. Lectins as cell recognition molecules. *Science,* 1989, N 246, p.227-234.

[10] Kellogg R.M. In:.*Host Guest Complex Chemistry Macrocycles.* Vögtle, F. Weber (Eds.) Springer-Verlag, Berlin, 1985. Chpt 7.

[11] Breslow R. Enzyme models related to inclusion compounds. In: Atwood J.L., Davies J.E.D., MacNicol D.D. (eds.). *Inclusion compounds.* Academic Press, London, 1984, v.3, p.473-508.

[12] Robertson A., Shinkai S Cooperative binding in selective sensors, catalysts and actuators. *Coord. Chem. Rev.*, 2000, v.205, N 1, 157-199.

[13] Rusin O, Král V.1,1'-Binaphtyl subunits as recognition groups of novel macrocyclic ligands. Application for saccharide sensing. *Sensors and Actuators B.*, 2001, v. 76, p. 331-335.

[14] Bell T.W., Hext N.M. Supramolecular optical chemosensors for organic analytes. *Chem. Soc. Rev.*, 2004, v.33, N 9, p.589-598.

[15] .Tbeur N, Rhlalou T., Hlaïbi M., .Langevin D., Métayer M., Verchère J.-F. Molecular recognition of carbohydrates by a resorcinarene. Selective transport of aditols through a supported liquid membrane. *Carbohydr. Res,* 2000, v. 329, p.409-422.

[16] .Berthod A., Chang S.S.C., Kullman J.P.S, Armstrong D.W. Practice and mechanism of HPLC oligosaccharide separation with a cyclodextrin bonded phase. *Talanta,* v. 47, p.1001-1012.

[17] Berezin B.D. *Porphyrins: Spectroscopy, electrochemistry, application.* Nauka, Moscow, Enikolopyan M.S. (ed) 1987. (in Russian)

[18] Lee J.D., Greene N.T., Rushton G.T., G.T..Shimizu G.T., Hong J.L. Carbohydrate recognition by porphyrin- based moleculary imprinted polymers. *Org.Lett.*, 2005, v.7, N 6, p. 963-966.

[19] Král V., Rusin O., Charvátová J., Anzenbacher P. Jr., and Fogl J.. Porphyrin phosphonates:novel anionic receptors for saccharide recognition. *Tetrahedr. Lett.*, 2000, v. 41, N 51, p.10147-10151.

[20] Charvátová J., Rusin O., Král V., Volka K., Matějka P.. Novel porphyrin based receptors for saccharide recognition in water. *Sensors and Actuators B.*, 2001, v. 76, p. 366-372.

[21] Hirata O., Yamamoto M., Sugiyasu K., Kobe Y., Ikeda M., Takeuchi M., Shinkai S. Allosteric saccharide sensing by a phenylboronic acids-appended 5,15-Bis (triarylethynyl)porphyrin. *J. Supramol. Chem.*, 2002, v. 2, N 1-3, p.133-142.

[22] Shinkai S., Ikedo M., Takeuchi M., Koumoto K. and Shinkai S. The first example of positive allosterism in aqueous saccharide-binding system designed on a Ce(IV)Bis(porphyrinate) double Decker scaffold. *Tetrahedron,* 2000, v.56, N 27, p.4717-4723.

[23] Hirata O., Kubo Y., Takeuchi M., Shinkai S. Mono- and oligosaccharide sensing by phenylboronic acid-appended 5,15-bis(diarylethynyl)porphoron complexes. *Tetrahedron,* 2004, v. 60, N 49, p. 11211-11218.

[24] Murakami H., Nagasaki T., Hamachi I., Shinkai S. Sugar sensing utilizing aggregation properties of a boronic-acid-appended porphyrin Tetrahedr. *Lett.,* 1993, v. 34, N 39,` P.6273-6276.

[25] Tong A.J., Yamauchi A., Hayashi T., Zhang Z.Y.,.Smith B.D, Teramae N. Boronic acid fluorophore/beta-cyclodextrin complex sensor for selective sugar recognition in water. *Anal. Chem.,* 2001, v. 73, N 7, p.1530-15306.

[26] S. Arimori, M. Takeuchi, S. Shinkai. Oriented molecular aggregates of porphyrin based amphiphiles and their morphology control by a boronic acid sugar interaction. *Supramol. Sci.,* 1998, v.5, N 1-2 . P. 1-8

[27] O. Rusin, V. Král. Novel macrocycles with 1,1'-binaphtyl substituents for the recognition of saccharides. *Chem. Commun.,* 1999, N 23, p. 2367-2368.

[28] K. Ladomenou, R.P. Bonar-Law. Urea porphyrins as a simple receptors for sugars. Chem. Commun. (Cambr.), 2002, p. 2108-2109.

[29] M. Dukh, D. Saman, K. Lang, V. Pouzar, I, Černy, P.Drašar, V. Král Steroid-porphyrin conjugates for saccharide sensing in protic media. *Org. Biomol. Chem.,* 2003, v 1, N 19, p.3458-3462.

[30] V. Král, O.Rusin, F.P. Schmidtchen Novel porphyrin-cryptand cyclic systems: receptors for saccharide recognition in water. *Org. Lett.,* 2001, v.3, N 6, p.873-876.

[31] Kobayashi N. Dimers, trimers and oligomers of phthalocyanines and related compounds. *Coord. Chem. Rev.,* 2002, v.227, p. 129-152.

[32] R.F. Pasternack, P.R. Huber, I. Boyd, G. Engasser, L. Francesconi, C. Gibbs, P. Fasella, G.C. Venturo, L. de C. Hind. On the aggregation of meso-substituted water-soluble porphyrins. *J. Am. Chem. Soc.,* 1972, v. 94, N 13, p. 4511-4517.

[33] W.B.Lu, L.H.Zhang, X.S.Ye . Porphyrin dimmers as receptors for selective binding of oligosaccharides. *Sensors and Actuators B,* 2006, v.113, N 1, p.354-360

[34] T. Mizutani, T. Kurahashi, T. Murakami, N.Matsumi, and H. Ogoshi. Molecular recognition of carbohydrates by zinc porphyrins: Lewis acid/Lewis base combination as s dominant factor for their selectivity. *J. Am. Chem. Soc.,* 1997, v.119, N 38, p.8991-9001.

[35] R.P.Bonar-Law, J.K.M. Sanders. Polyol recognition by a steroid-capped porphyrin. Enhancement and modulation of misfit guest binding by added water or methanol. *J.Am. Chem. Soc.,* 1995, v.117, N 1, p. 259-271.

[36] J.D.Watson. *Molecular Biology of Gene,* 3rd ed., Benjamin, New York, 1976, p.98.

[37] Singelenberg F.A.J, Van der Maas J.H. Hydrogen bonding in non cyclic vicinal diols and their mono-methyl ethers: an ftir study. *J. Mol. Struct.,* 1991, v. 243, N 1-2, p. 111 -122.

[38] Kurahashi T., Misutami T., Yoshida Jun-ichi. Effect of intramolecular hydrogen-bonding network on the relative reactions of carbohydrate OH-groups. *J. Chem Soc. Perkin Trans.* 1, 1999, v. 999, N 4, p.465-474.

[39] M. Inouye, T. Miyake, M.Furusyo, and H. Nakazumi. Molecular recognition of β-ribofuranosides by synthetic polypyridine-macrocyclic receptors. *J. Am. Chem. Soc.,* 1995, v. 117, N 50, p.2416-12425

[40] Y. Aoyama, Y. Tanaka, and S. Sugahara Molecular recognition. 5. Molecular Recognition of Sugars via Hydrogen-Bonding Interaction with Synthetic Polyhydroxy Macrocycle. *J. Am. Chem. Soc.,* 1989, v. 111, N 14, p. 5397-5404.

[41] K. Kurihara, K. Ohto, Y, Tanaka, Y, Aoyama, and T. Kunitake Molecular recognition of sugars by monolayers of resorcinol-dodecanal cyclotetramer. *J. Am. Chem. Soc.,* 1991, v. 113, N 2, p. 444-450.

[42] Y. Kikuchi, Y. Tanaka, S. Sutarto, K. Kobayashi, H.Toi, and Y. Aoyama. Highly cooperative binding of alkyl glucopyranosides to the resorcinol cyclic tetramer due to intracomplex guest-guest hydrogen-bonding: Solvophobicity/solvophilicity control by an alkyl group of the geometry, stoichiometry, stereoselectivity, and cooperativity. *J. Am. Chem. Soc.,* 1992, v. 114, N 26, p. 10302-10306.

[43] O. Rusin, V. Král. 1,1'-Binaphtyl substituted calixresorcinols-methyl red complexes: receptors for optical saccharide sensors. *Tetrahedr. Lett.,* 2001, v. 42, p .4235-4238.

[44] K. Kobayashi, Y. Asakawa, Y. Kato, and Y. Aoyama. Complexation of hydrophobic sugars and nucleosides in water with tetrasulfonate derivatives of resorcinol cyclic tetramer having a polyxydroxy aromatic cavity: Importance of guest-host CH-π interaction. *J. Am. Chem. Soc.,* 1992, v. 114, N 26, p. 10307-10313.

[45] Miyajima K., Machida K., Taga T., Komatsu H. and Nakagaki M. Correlation between the hydrophobic nature of monosaccharides and cholates and their hydrophobic indices. *J. Chem. Faraday Trans.* I, 1988, v.84, N 8, p. 2537-2544.

[46] Lemieux R.U. The origin of the specificity in the recognition of oligosaccharides by proteins. *Chem. Soc. Rev.,* 1989, v.18, p.347-374.

[47] Yanado M., Yano Y. Hydrophobic nature of sugars as evidenced by their differential affinity for polystyrene gel in aqueous media. *J. Solut. Chem.,* 1985, v.14, N 12, p.891-902.

[48] Goodnow T.T., Reddington M.V., Stoddartand J.F., Kaifer A.E. Cyclobis(paraquat-p-phenylene0: A novel synthetic receptor for amino acids with electron-rich aromatic moieties. *J. Am. Chem. Soc.,* 1991, v.113, N 11, p. 4335-4337.

[49] Staley S.A and Smith B.D.. Phenyl glycopyranoside recognition in water using Stoddart's cyclobis(paraquat-p-phenylene) receptor. *Tetraherd. Lett.,* 1996, v. 37, N. 3, p. 283-286.

[50] Lipton M.A. Diastereoselective recognition of β anomers of phenyl glycosides by a cyclobis(paraquat-p-phenylene) receptor: A computational study. *Terahedr. Lett.,* 1996, v.37, N 3, p. 287-290.

[51] Breslow R., Dong S.D. Biomimectic reactions catalysed by cyclodextrin and their derivatives. *Chem. Rev.,* 1998, v. 98, N 5, p.1997-2011.

[52] Szejtli J. *Cyclodextrin Technology*. Dordrecht: Kluwer Academic, 1988.

[53] Singh M., Sharma R., Banerjee U.C. Biotechnological applications of cyclodextrins. *Biotechnol Adv.*, 2002, v. 20, p.341-359.

[54] Shpigun O.A., Ananieva I.A., Budanova N.Yu., Shapovalova E.N. The use of cyclodextrins for separation of enantiomers. *Russ. Chem. Rev.*, 2003, v. 72, N 12, p. 1167-1180.

[55] Danil de Namor A. F., Blackett P.M., Cabaleiro M.C., Al Raw J.M.A. Cyclodextrin-monosaccharide interactions in water. *J. Chem. Soc. Faraday Trans. 1.*. 1994., v .90, N 6, p. 845-847.

[56] Eliseev A.V., Schneider H.-J. Molecular recognition of nucleotides, nucleosides, and sugars by aminocyclodexrins. *J. Amer. Chem. Soc.*, 1994, v. 116, N. 14, p. 6081-6088.

[57] Ternai M., Buvári-Barcza Á and Barcza L. Dual behaviour of maltosyl-β-cyclodextrins: investigation of formation of inclusion and glucose-type complexes. *J. Inclus. Phenom. Macrc. Chem.*, 1999, v. 34, p.311-319.

[58] Armstrong D.W. and Jin H.L. Evaluation f the liquid chromatographic separation of monosaccharides, disaccharides, trisaccharides, tetrasaccharides, deoxysaccharides and sugar alcohols with stable cyclodextrin bonded phase columns. *J. Chromatogr.*, 1989, v. 462, p. 219-232.

[59] Berthod A., Chang S.S.C., Kullman J.P.S., Armstrong D.A. Practice and mechanism of HPLC oligosaccharide separation with a cyclodextrin bonded phase. *Talanta,* 1998, v. 47, N 4, p. 1001-1012.

[60] Bakshi M.S. Host-guest interactions. I. Volumes of the sucrose+β-cyclodextrin+water ternary systems at 25^0C. *J. Solut. Chem.*, 1996, v. 25, N 4, p. 411-420.

[61] .Jiménez-Barbero J., Junquera E., Martín-Pastor M., Sharma S., Vicent C., and Penadés S. Molecular recognition of carbohydrates using synthetic receptor. A model system to understand the stereoselectivity of a carbohydrate-carbohydrate interaction in water. *J. Am. Chem. Soc.*, 1995, v.117, N 45, p.11198-11204.

[62] Coterón J.M., Vicent C., Bosso C., and Penadés S. Glycophanes, cyclodextrin-cyclophane hybrid receptors for apolar binding in aqueous solutions. A stereoselective carbohydrate-carbohydrate interaction in water. *J. Am. Chem. Soc.*, 1993, v. 115, N 22, p.10066-10076.

[63] Morales J.C. and Penadés S. Two interconverting glycophanes from maltose. *Tetrahedr. Lett.*, 1996, v.37, N28, p.5011-5014.

[64] Hiraoka M. *Crown Compounds. Their Characteristics and Applications.* Elsevier, Amsterdam, 1982.

[65] Izatt R.M., Bradshaw J.S., Nielsen S.A., Lamb J.D., Christensen J.J., Sen D. Thermodynamic and kinetic data for cation-macrocycle interactions. *Chem. Rev.*, 1985, v.85, N 4, p.271-274.

[66] Parfenyuk E.V., Kulikov O.V. Enthalpic and volume effects in the interaction of D-glucose with crown ethers in water. *Thermochim. Acta,* 1996, v. 285, N. 2, 253-258.

[67] Parfenyuk E.V. Heats of interaction of D-maltose with crown ethers in water. *Russ. J. Phys. Chem.*, 1998, v. 72, N. 11, 2102-2103.

[68] Parfenyuk E.V. Peculiarities of interactions of 18-crown-6-ether with mono- and disaccharides in water and dimethylformamide. *Russ. J. Phys. Chem.*, 2001, v. 75, N 9, p.16085-1610.

[69] Davydova O.I, Parfenyuk E.V. Calorimetric study of the interactions of disaccharides with crown ethers in water. *Mendeleev Commun.*,2001, N 5, p. 203-204.

[70] Davydova O.I., Lebedeva N.Sh., Parfenyuk E.V. Calorimetric study of binding of some disaccharides with crown ethers. *Thermochim. Acta,* 2004, v.421, N 1, p.31-33.

[71] Parfenyuk E.V., Lebedeva N.Sh. Effect of type of glycosidic linkage on selective interactions of maltose and cellobiose with some crown ethers in dilute solution. *J. Solut. Chem.* 2007, v. 36, N 1, p. 95-107.

[72] Parfenyuk E..V., Davydova O.I. Molecular complexes of mono- and oligosaccharides in solutions. In: *Chemistry of Polysaccharides.* (Ed.G.E.Zaikov.) Brill, Nertherlands, 2005. Chpt.1.

[73] Volkova N.L., Parfenyuk E.V. Selective interactions of 18-crown-6 with D-glucose and D-galactose in aqueous solutions: Titration calorimetry, densimetry, viscosimetry. Thermochim. *Acta,* 2005, v.431, N 1, 108-112.

[74] Volkova N.L., Parfenyuk E.V. Parameters of molecular complex formation in ternary solutions D-glucose or D-galactose – 15-crown-5-ether -water . *Russ. J. Phys. Chem.* 2007, v.81, (in press)

[75] Volkova N.L., Lebedeva N.Sh. Parfenyuk E.V. Complexation of D-xylose and L-arabinose with 18-crown-6 in aqueous solutions: Calorimetric, densimetric, and vicsosimetric studies. *Russ. J. Coord. Chem.,* 2005, v. 31, N 12, p. 899-903

[76] Volkova N.L., Barannikov V.P., Parfenyuk E.V. Thermochemical behavior of 18-crown-6 in aqueous solutions. *Thermochim. Acta,* 2006, v.444, N 1. 13-15.

[77] Fukuhara K., Tachikake M., Matsumoto S., Matsuura H. Raman spectroscopic study of the hydrates of 18-crown-6. *J. Phys. Chem.,* 1995, v. 99, N 21, p. 8617-8623.

[78] Fenau-Dupont J., Declercq J.P., Germain G., Mursche M.V. Complexe 1 : 1 pentaoxa-1,4,7,10,13-cyclopentadécane-bromure cuire(II) hydrate. *Acta Crystallogr.,* 1979, v. .B35, N 5, p. 1215-1217.

[79] Brinnger L.-E., Wadsö I. Some thermodynamic properties of crown ethers in aqueous solution. *J. Chem. Thermodyn.,* 1990, v. 22, p.143-148.

[80] Conner A.H. and Anderson L. The tautomerization and mutarotation of β-L-arabinose. Participation of both furanose anomers. *Carbohydr. Res.* 1972, v. 25, N 1, p. 107 -116.

[81] Pérez S. *Sucrose properties and applications.* Eds. M. Mathlouthi, P.H.Reiser, Chapman and Hall, London, 1994, p.11.

[82] Kačuráková M., Mathlouthi M. FTIR and laser-Raman spectra of oligosaccharides in water: characterization of the glycosidic bond. *Carbohydr. Res,* 1996, v. 248, N 2, p.145-157.

[83] Magazù S., Migliardo P., Musolino A.M., and Sciortino M.T. α,α-Trehalose-water solutions. 1. Hydration phenomena and anomalies in the acoustic properties. *J. Phys. Chem. B.,* 1997, v.,101, N 13, p. 2348-2351.

[84] Branca C., Magazù S., Maisano G. and Migliardo P. Experimental study of the hydration properties of homologous disaccharides. J. Biol. Phys., 2000, v. 26, 295-306.

[85] Almond A. Towards understanding the interaction between oligosaccharide and water molecules. *Carbohydr. Res.,* 2005, v. 340, N 5, p. 907-920.

[86] Bakshi M.S. host-guest interactions: Part 2. Volumes of β-cyclodextrin+18-crown-6 ether+water and sucrose+18-crown-6 ether+ water ternary systems at 25^0 C. *Indian J. Chem.,* 1997, v. 36A, N 11, p. 931-937.

[87] Bonar-Law R.P., Davis A.P. Cholic acid as an architectural component in biomimetic/molecular recognition chemistry; synthesis of the first "cholophanes". *Tetrahedron,* 1993, v. 49, N 43, p. 9845-9854.

[88] Evans S.M., Burrows C.J., Venanzi C.A. Design of cholic acid macrocycles as hosts for molecular recognition of monosaccharides. *J. Mol. Struct.*(Theochem),1995, v. 334, p. 193-205.

[89] Niidome T., Murakami H., Kawazoe M., Hatakeyama T., Kobashigawa Y., Matsushita M., Kumaki Y., Demura M., Nitta K. and Aoyagi H. Carbohydrate recognition of gramicidin S analogues in aqueous medium. *Bioorg. Med.Chem. Lett.,* 2001, v. 11,p. 1893-1896.

In: Solution Chemistry Research Progress
Editor: Darian V. Bostrelli, pp. 93-113
ISBN 978-1-60021-993-1

Chapter 4

SELECTIVE INTERACTIONS OF NATIVE AND MODIFIED CYCLODEXTRINS WITH SOME B-VITAMINS IN AQUEOUS SOLUTION. THERMODYNAMIC STUDY

Irina V. Terekhova*
Institute of Solution Chemistry of Russian Academy of Sciences
1 Akademicheskaya str., 153045 Ivanovo, Russian Federation

ABSTRACT

To modify the physical-chemical and biological properties of biomolecules and to prolong their biological action, the ability of native and hydroxypropylated α- and β-cyclodextrins to form inclusion complexes with some B-vitamins (pyridoxine, pyridoxal, nicotinamide, nicotinic acid) in aqueous solution at 298.15 K has been studied by means of calorimetry, capillary electrophoresis and ^1H NMR spectroscopy. Two thermodynamic approaches were applied to characterize the complexation processes and weak interactions that are not accompanied by complex formation. In the first case, the thermodynamic properties of complexation such as K, Δ_cG, Δ_cH and Δ_cS were considered. For interpretation of weak interactions, the enthalpic virial coefficients (h_{xy}) were calculated according to the McMillan-Mayer theory. On the basis of the experimental data, it was found that cyclodextrins exhibit selectivity in the interactions with B-vitamins. The formation of 1:1 enthalpy stabilized complexes was observed in the systems (α-cyclodextrin + nicotinic acid), (hydroxypropyl-α-cyclodextrin + nicotinic acid), (α-cyclodextrin + isonicotinic acid), (α-cyclodextrin + nicotinamide) and (β-cyclodextrin + nicotinamide). The obtained thermodynamic parameters of interactions were discussed in terms of influence of the following factors: the structure of guest molecules; availability of the macrocyclic cavity and its dimensions; the presence of

* E-mail: ivt@isc-ras.ru

hydroxypropyl substituents and medium pH. Using the ^1H NMR data the possible modes of inclusion of guest molecule into cyclodextrin cavity were proposed.

Keywords: Cyclodextrin; B-vitamin; complex formation; thermodynamics

INTRODUCTION

Cyclodextrins (CD) are a family of macrocyclic receptors, which are able to form inclusion complexes with a wide variety of organic compounds [1-3]. Cyclodextrins are cyclooligosaccharides obtained by the enzymatic disintegration of starch. The well-known α-, β- and γ-cyclodextrins consist of six, seven and eight D-glucopyranose units, respectively, linked by α-glucosidic bonds. Cyclodextrins have a well-defined structure which can be described as a truncated cone with a hydrophobic cavity suitable for inclusion of guest molecules of appropriate size and shape. Hydroxyl groups surrounding the narrow and wide rims of the cyclodextrin molecule make the cyclodextrin external surface hydrophilic and inclusion complexes water soluble. It was established that the forces involved in the inclusion complex formation are of non-covalent nature (hydrophobic, van der Waals and electrostatic interactions as well as hydrogen bonding) [1-4]. A variety of cyclodextrin derivatives, in which the OH-groups were substituted by hydroxypropyl, methyl-, carboxymethyl- and other different groups, was synthesized recently [5, 6]. Cavity dimensions, hydrophobicity and flexibility of the derivatized cyclodextrins can be noticeably modified. Moreover, the substituted cyclodextrins can display higher selectivity and form more stable inclusion complexes due to the additional interactions of their functional substituents with the guest molecules.

Almost all industrial applications of cyclodextrins are based on their ability to form inclusion complexes. Cyclodextrins are considered and used as encapsulating materials since they are the naturally occurring oligosaccharides and, therefore, nontoxic. The main advantages of the application of cyclodextrins in pharmaceutical, cosmetic, food and textile industries are as follows: stabilization of labile compounds, enhancement of solubility and bioavailability, prolongation of biological and pharmacological activity, reduction or elimination of undesired physical-chemical properties (taste, odour) and side effects of drugs [7-10]. Owing to chiral structure and selective complexation with isomeric compounds cyclodextrins can be employed in such separation techniques as chromatography and capillary electrophoresis [10-12].

The aim of our work is to reveal the ability of native and substituted cyclodextrins to form inclusion complexes with biologically active guest molecules in aqueous solution. Such B-vitamins as nicotinic acid, nicotinamide, pyridoxal and pyridoxine (Figure 1) were chosen as guest molecules. They are pyridine derivatives and have the similar structure discriminated by the nature and number of side groups. These vitamins are necessary for normal functioning of a living organism. Furthermore, they are important compounds in pharmaceutical, cosmetic and food products. For instance, nicotinic acid (vitamin B_5) is used as an antihyperlipidemic agent and reduces the cholesterol level [13]. Nicotinamide may have antidiabetogenic, antioxidant, antiinflammatory and anticarcinogenic activities. Pyridoxine and pyridoxal (B_6 vitamins) are required for the balancing of hormonal changes as well as assisting the immune and nervous systems. Besides the numerous positive effects these vitamins have a number of disadvantages. Flushing and hot flashes are known as unwanted side effects of the nicotinic acid [14]. Vitamin B_6 is unstable under UV-radiation. Moreover, activity of vitamins can be considerably reduced due to the oxidation and possible interactions with chemicals during the storage and processing. To prolong the pharmacological action of vitamins and to counteract their unpleasant properties, the encapsulation by cyclodextrins can be proposed and used. In addition, complex formation of vitamins with cyclodextrins can be employed for separation of vitamins and determination of their quantity in multicomponent mixtures or pharmaceutical formulations.

Only a few studies on complex formation of cyclodextrins with pyridine [15] and pyridine derivatives [16-18] have been published. The stability constant and thermodynamic parameters of 1:1 complexation of α-cyclodextrin with pyridine in aqueous solution have been determined by Lewis and Hansen [15]. These thermodynamic quantities will be presented and discussed below. As it was shown in [16], nicotinic acid fully ionized at pH 7.0 and methyl nicotinate display low binding affinity to β-cyclodextrin in aqueous solution. The crystalline 1:1 and 1:2 complexes of β-cyclodextrin with 1,4-dihydropyridine derivatives were isolated and characterized by X-ray, DSC and ^1H NMR methods [18]. The inclusion of aromatic fragments of these guests within the cyclodextrin cavity was elucidated by means of ^1H NMR chemical shift measurements. Thus, no studies on the interactions of cyclodextrins with B-vitamins under consideration have been published with the exception of [16] and our works [19-22]. The present paper summarizes the results of calorimetric (using home-made isothermal calorimeter of solution [23] and isothermal titration microcalorimeter Omega MicroCal Inc.), ^1H NMR (Bruker AC-200 spectrometer operating at 200 MHz) and capillary electrophoresis (Beckman P/ACE 5510 instrument, Germany) studies on complex formation of native and modified cyclodextrins with some B-vitamins in aqueous solution at 298.15 K. Influence of the vitamin structure, availability of the macrocyclic cavity and its dimensions, the presence of hydroxypropyl substituents in the cyclodextrin rims and the medium pH on the thermodynamics of interaction and binding mode is analyzed.

Results and Discussion

Interactions of Cyclodextrins with Nicotinic acid and Isonicotinic Acid

We started the investigation of interactions of cyclodextrins with B-vitamins from the consideration of nicotinic acid (3-pyridinecarboxylic acid) containing a minimal number of substituents in the pyridine ring. Isonicotinic acid (4-pyridinecarboxylic acid) having no biological activity was also considered as a structural isomer of the nicotinic acid. As it is well known, in aqueous solution nicotinic and isonicotinic acids can exist as cation, zwitterion and anion. Distribution of the nicotinic acid species depending on pH is illustrated in Figure 2. Dependences similar to those shown in Figure 2 were plotted for the isonicotinic acid. Thus, cationic species predominate at $pH<2.0$, zwitterions – in the pH rage 3-4 and anions at $pH>6.5$.

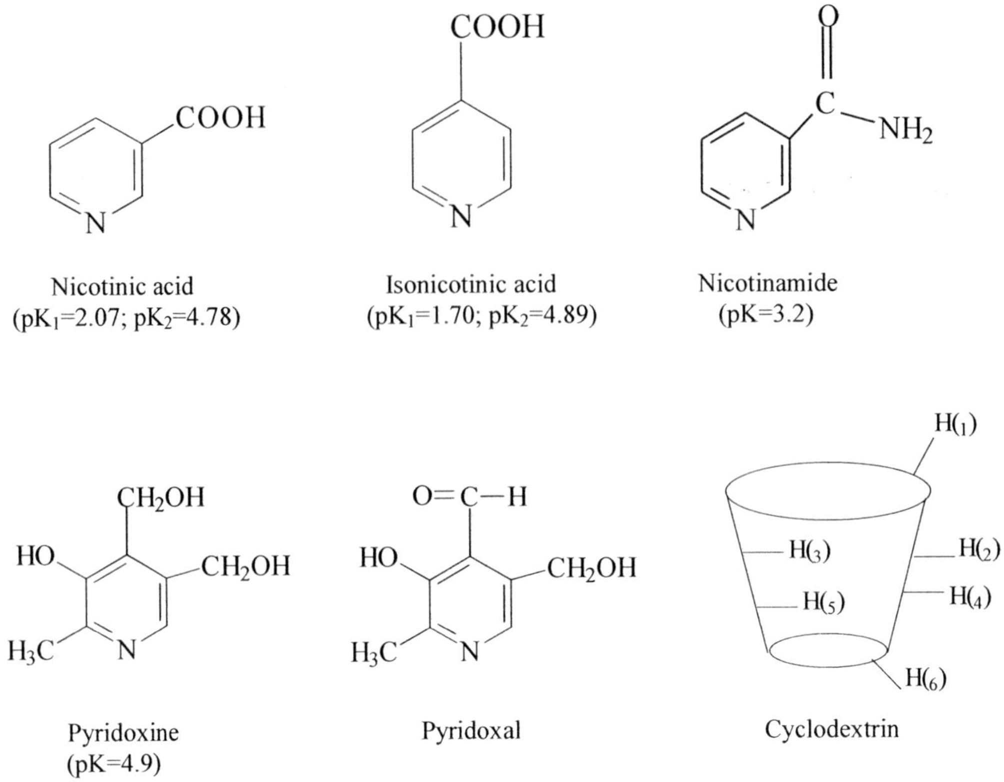

Figure 1. Objects under study.

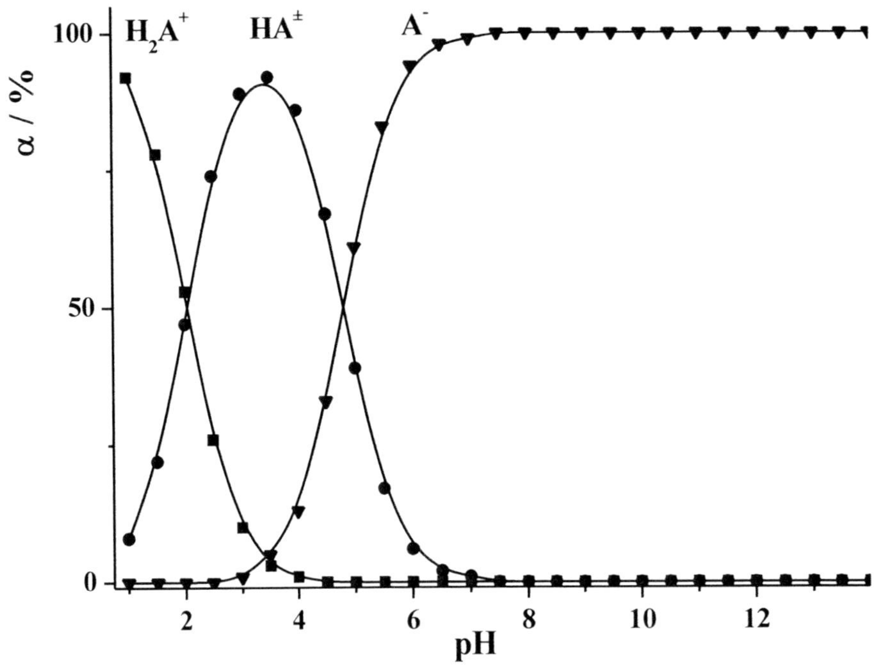

Figure 2. Distribution of nicotinic acid species depending on pH.

Figures 3 and 4 show the experimental dependences of enthalpy of transfer of cyclodextrins and its structural units (α-D-glucose and maltose) from water to aqueous solutions of nicotinic acid of the variable concentration. Different character of these dependences points out the different binding ability of α- and β-cyclodextrin to the nicotinic acid. As follows from the binding isotherms presented in Figure 3, nicotinic acid forms complexes only with α- and hydroxypropyl-α-cyclodextrin possessing the smaller cavity diameter. The linear dependences obtained for the interactions of nicotinic acid with β- and hydroxypropyl-β-cyclodextrin as well as with glucose and maltose (Figure 4) correspond to the weak interactions. Thus, availability of the macrocyclic cavity and its size play an important role in the binding of cyclodextrins with the nicotinic acid.

Table 1. Enthalpic coefficients of pair interactions of cyclodextrins with some vitamins in water at 298.15 K

System	pH	h_{xy} (J-kg-mol^{-2})
nicotinic acid + α-D-glucose	3.4	7.1 (±0.2)[a]
nicotinic acid + maltose	3.4	10.1 (±0.4)
nicotinic acid + β-CD	3.4	-1.6 (±0.1)
nicotinic acid + HP-β-CD	3.4	12.1 (±0.2)
nicotinic acid + HP-β-CD	6.8	69.2 (±0.8)
nicotinic acid + HP-α-CD	6.8	207.8 (±0.5)
isonicotinic acid + β-CD	3.4	-1.5 (±0.9)
nicotinamide + HP-α-CD	6.0	-9.7 (±0.4)
nicotinamide + HP-α-CD	2.0	-0.9 (±0.2)
pyridoxine + α-CD	6.8	82.3 (±1.2)
pyridoxine + HP-α-CD	6.8	78.9 (±0.5)
pyridoxine + β-CD	6.8	21.1 (±0.7)
pyridoxine + HP-β-CD	6.8	73.5 (±0.4)
pyridoxal + α-CD	6.8	137.4 (±1.5)
pyridoxal + β-CD	6.8	74.3 (±0.5)

[a] Standard deviation.

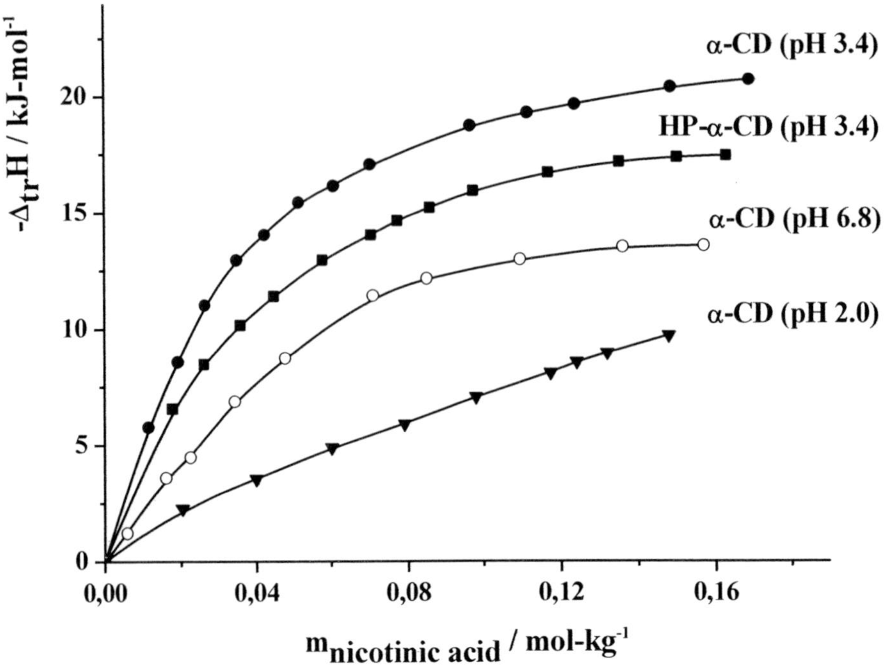

Figure 3. Enthalpy of transfer of cyclodextrins from water to aqueous solutions of nicotinic acid *versus* the nicotinic acid concentration (298.15 K).

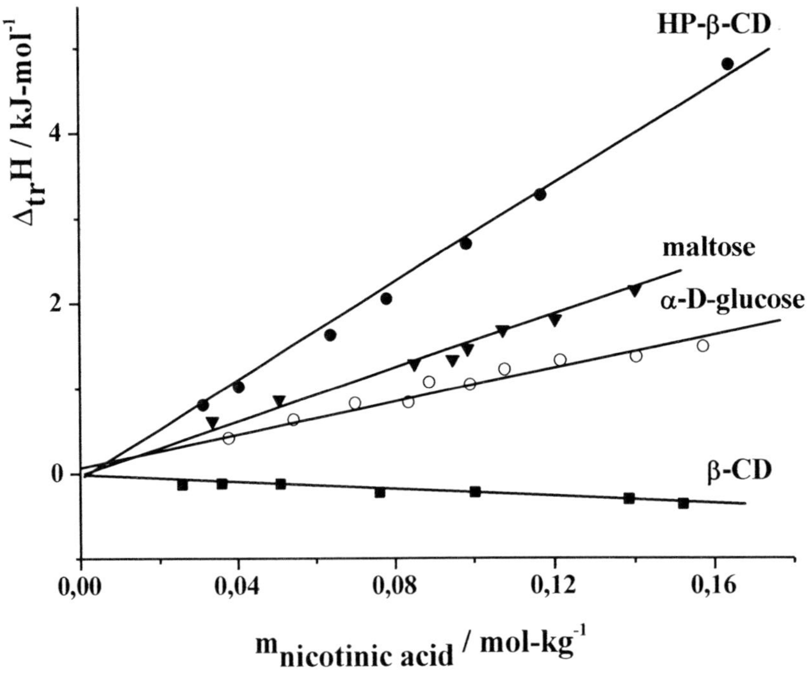

Figure 4. Enthalpy of transfer of saccharides from water to aqueous solutions of nicotinic acid *versus* the nicotinic acid concentration (298.15 K; pH 3.4).

Weak interactions were characterized by enthalpic virial coefficients calculated from the calorimetric measurements using McMillan-Mayer approach [24, 25]:

$$\Delta_{tr}H(w \rightarrow w+y) = 2h_{xy}m_y + 3h_{xyy}m_y^2 + 3h_{xxy}m_xm_y + \ldots \quad (1)$$

where m_x and m_y are the molalities of saccharide and guest compound, respectively; h_{xy}, h_{xyy} and h_{xxy} are the enthalpic coefficients of pair and triplet interactions. Since the saccharide concentration was very low (~$1 \cdot 10^{-3}$ mol-kg^{-1}) the last term in Equation (1) can be neglected. Thus, Equation (1) can be easily transformed into Equation (2):

$$\Delta_{tr}H(w \rightarrow w+y)/m_y = 2h_{xy} + 3h_{xyy}m_y \quad (2)$$

from which the h_{xy} can be calculated by the linear least squares analysis. The values of h_{xy} coefficients are reported in Table 1. The enthalpic virial coefficients reflect the total change in the enthalpy of the solution caused by solute-solute and solute-solvent interactions.

As can be seen from Table 1, for interactions of nicotinic acid with α-D-glucose and maltose the positive h_{xy} were obtained. It suggests that endothermic effects caused by the dehydtration of solutes dominate over specific interactions (e.g. van der Waals interactions, H-bonding). In comparison with glucose, the interaction of nicotinic acid with maltose becomes more enthalpically unfavorable due to the increased positive contribution from dehydration of maltose which is disaccharide consisting of two glucose units. To the contrary, the interaction of nicotinic acid with β-cyclodextrin consisting of seven glucose units is characterized by the negative h_{xy} (Table 1). Probably, availability of the macrocyclic cavity favors the attractive interactions, which determine the prevalence of exothermic effects. However, compensation of the negative enthalpy by the large negative entropy results in low free energy value and absence of stable complexes in this system.

As it was mentioned above, α-cyclodextrin consisting of six glucose units forms complexes with the nicotinic acid in aqueous solution. Apparently, binding takes place due to geometric complementarity of α-cyclodextrin cavity size to the structure of nicotinic acid molecule. Calculated thermodynamic parameters of complex formation (K, Δ_cG, Δ_cH and Δ_cS) are reported in Table 2. A good agreement between values obtained by the different experimental methods should be noted. As can be seen from Table 2, complexation of the nicotinic acid with α-cyclodextrin is characterized by high negative enthalpy and entropy values, which compensate each other resulting in low equilibrium constants. Thus, complexes are only enthalpy stabilized.

Enthalpy and entropy terms include the contributions from the following processes that take place during the binding [3, 4]:

1) dehydration of solutes;
2) conformational changes;
3) release of cavity bounded high-energy water and its replacement by guest molecules;
4) binging of cyclodextrin with guest through H-bonding, van der Waals, electrostatic and hydrophobic interactions;
5) hydration of complex.

Table 2. Thermodynamic parameters of complex formation of cyclodextrins with some guest molecules in water at 298.15 K

Complex	pH (kg-mol^{-1})	K (kJ-mol^{-1})	$\Delta_c G$ (kJ-mol^{-1})	$\Delta_c H$ (kJ-mol^{-1})	$T\Delta_c S$	Method
α-CD/nicotinic acid	3.4	33(±5)	-8.7	-26.4 (±0.3)	-17.7	cal
		30 (±2)	-8.4	-24.5(±3.5)	-16.1	CE
	2.0	3.2(±0.6)	—	—	—	CE
	6.8	14(±2)	-6.5	-23.5 (±0.6)	-17.0	cal
HP-α-CD/nicotinic acid	3.4	23(±4)	-7.8	-24.3 (±0.3)	-16.5	cal
		17 (±1)	—	—	—	CE
	2.0	no complex formation				CE
	6.8	no complex formation				cal
α-CD/isonicotinic acid	3.4	7 (±2)	-4.8	-13.4 (±0.4)	-8.6	cal
		12 (±6)	—	—	—	CE
	2.0	4 (±2)	—	—	—	CE
HP-α-CD/isonicotinic acid	3.4	No complex formation				CE
	2.0	No complex formation				CE
α-CD/nicotinamide	6.0	4(±2)	-3.4	-6.6(±0.4)	-3.2	cal
	2.0	No complex formation				CE
β-CD/nicotinamide	6.0	8(±3)	-5.2	-9.8 (±0.5)	-4.6	cal
	2.0	No complex formation				CE
α-CD/pyridine[15]		158 (±101)	-12.6	-10.5 (±3.8)	2.1	cal
α-CD/benzoic acid [15]		1000 (±33)	-17.1	40.2 (±0.4)	-22.5	cal

Cal – calorimetry; CE – capillary electrophoresis.

Since $\Delta_c H$ and $\Delta_c S$ are high and negative, the binding of the nicotinic acid with α-cyclodextrin is realized primarily due to the van der Waals and electrostatic interactions as well as the H-bonding. To confirm this assumption the comparison of the thermodynamic parameters of α-cyclodextrin complex formation with pyridine [15], benzoic acid [15] and nicotinic acid is necessary (Table 2).

As follows from Table 2, most stable complexes are formed between α-cyclodextrin and benzoic acid. Binding is accompanied by the highest negative enthalpy and entropy values which can be attributed to the prevalence of attractive van der Waals interactions and H-bonding as well as to the conformational restriction of both host and guest. Since possible hydrophobic interactions of the benzene ring with the cyclodextrin cavity should give the positive contribution to $\Delta_c H$ and $\Delta_c S$, one can assume the important role of the carboxylic group in the binding of benzoic acid with α-cyclodextrin. Appearance of the nitrogen atom in the aromatic ring of the nicotinic acid molecule results in the considerable weakening of binding and decreasing of $\Delta_c H$ and $\Delta_c S$. Probably, the presence of the nitrogen in the aromatic ring influences the hydration state of the nicotinic acid molecule. In particular, the hydration of the nicotinic acid becomes stronger in comparison with the benzoic acid and, as consequence, prevents the binding with α-cyclodextrin. In contrast to the nicotinic and benzoic acids, pyridine forms with α-cyclodextrin enthalpy-entropy stabilized complex. It may be, that the absence of the carboxylic group changes the binding mode of pyridine with α-cyclodextrin. Increasing of Δ_cH and Δ_cS values can be a result of deeper penetration of the pyridine into macrocyclic cavity and prevalence of the hydrophobic interactions and more extensive dehydration.

Comparison of the binding ability of α-cyclodextrin to the different nicotinic acid species allows to emphasize the following. Complex formation of α-cyclodextrin with the zwitterionic and anionic species having the ionized carboxylic group is characterized by the approximately identical thermodynamic parameters (Table 2). Minor decrease of exothermicity of binding with the anionic form in comparison with the zwitterionic form can be explained by the deprotonation of nitrogen. Perhaps, the protonated nitrogen of zwitterion carrying the positive charge can participate in the interactions with α-cyclodextrin and additionally stabilize the complex. On the contrary, protonation of the carboxylic group results in considerable decreasing of stability of complex formed by cationic species and α-cyclodextrin. Unfortunately, correct calculation of the Δ_cH and Δ_cS values for complex formation of cationic form of the nicotinic acid with α-cyclodextrin was problematic due to very low equilibrium constant and high inaccuracy of their determination. Thus, on the basis of the obtained results it is possible to assume that availability of the ionized carboxylic group favors the complex formation of the nicotinic acid with α-cyclodextrin. The same tendency is observed for the complex formation of α-cyclodextrin with isonicotinic acid.

Comparative analysis of complex formation of α-cyclodextrin with nicotinic and isonicotinic acids at pH 3.4 shows that position of the carboxylic group versus the nitrogen in the pyridine ring significantly influences the thermodynamic parameters (Table 2). Replacement of the carboxylic group in *para*-position of the aromatic ring results in less enthalpically favorable binding and formation of less stable complex. The difference in the binding ability of these isomeric acids can be explained by the following facts known from literature. The electronic charge distribution in the pyridinecarboxylic acid molecule is changed depending on the position of the carboxylic group versus the nitrogen [26]. Moreover, the hydration state, acid-base properties and capability to complexation also depend on the location of carboxylic group versus the nitrogen in the pyridine ring. Additionally, it should be mentioned that nicotinic acid can exist in two conformations which differ in the spatial location of the carboxylic group, whereas only one stable conformer of isonicotinic acid was found [26]. Consequently, placing the negatively charged carboxylic group closer to the positively charged nitrogen atom is favorable for binding with α-cyclodextrin. The same conclusion can be drawn for complex formation of hydroxypropylated α-cyclodextrin with the pyridinecarboxylic acids under study.

It is interesting to compare the binding ability of native and modified cyclodextrins towards the pyridinecarboxylic acids. In some cases, substitution of OH-groups surrounding the cyclodextrin cavity by the functional groups can cause the increase of depth and diameter of the cavity as well as the change of hydrophobicity and flexibility promoting more effective binding. However, the opposite influence of substituents when they prevent the complex formation by creating of steric hindrance to the inclusion of the guest into the host cavity is also known. As it is evident from Table 2, the partial substitution of OH-groups surrounding the narrow and wide rims of α-cyclodextrin molecule (average substitution degree of hydroxypropyl-α-cyclodextrin was 0.6 per glucopyranose unit) results in a decrease of the stability constants of the complexes formed. These bulky substituents can spatially obstruct the binding of nicotinic acid with hydroxypropyl-α-cyclodextrin, and in the case with isonicotinic acid prevent the complex formation process. Steric effects may be responsible for this phenomenon. Additionally, the increased positive contribution from dehydration of bulky hydroxypropyl groups can play a significant role. As it is evident from Table 2, the complexation process of hydroxypropyl-α-cyclodextrin with nicotinic acid becomes less enthalpically favorable and more entropically favorable. The same tendency is observed for interactions of nicotinic acid with hydroxypropyl-β-cyclodextrin which are accompanied by higher endothermic effects as compared with unsubstituted β-cyclodextrin (Table 1).

To obtain information on binding mode the carrying out of ^{1}H NMR measurements can be useful. For this purpose, the ^{1}H NMR spectra of nicotinic and isonicotinic acids in the presence of excess amounts of α- and β-cyclodextrin were recorded. The changes in the chemical shifts of protons are listed in Table 3. As follows from Table 3, the signals from H(2), H(4) and H(5) protons of nicotinic acid are more affected by the addition of α-cyclodextrin. It means that the part of the nicotinic acid molecule where carboxylic group is placed participates in the binding with α-cyclodextrin. This conclusion is in accordance with the binding mode of benzoic acid and its derivatives with α-cyclodextrin in D_2O solutions [27-31]. For β-cyclodextrin/nicotinic acid complexes, the measurable $\Delta\delta$ observed for H(4) and H(5) suggest the shallow insertion of the nicotinic acid aromatic ring into β-cyclodextrin cavity. The small $\Delta\delta$ values obtained for isonicotinic acid protons (Table 3) correlate with the lower binding affinity of cyclodextrins to this acid.

To characterize the nature of the complexes and the depth of guest penetration into cyclodextrin cavity the consideration of ^{1}H NMR spectra of α- and β-cyclodextrin in the presence on nicotinic and isonicotinic acids is necessary. It is well known from literature [31], that ^{1}H NMR spectra of α- and β-cyclodextrin consist of signals from H(1), H(2), H(4) and H(6) protons located in the exterior of the molecule, and H(3) and H(5) protons placed inside the cavity at the wider and narrower rims, respectively. Namely, H(3) and H(5) protons are sensitive to inclusion of guest molecule into hydrophobic cavity.

Table 3. Chemical shift changes of pyridinecarboxylic acid protons induced by the addition of cyclodextrins (pH=3.4, D_2O, 298.15 K)

	$\Delta\delta^*$ (ppm)					
CD	nicotinic acid (0.005 mol-kg^{-1})				isonicotinic acid (0.005mol-kg^{-1})	
	H(2)	H(4)	H(5)	H(6)	H(2),(6)	H(3),(5)
α-CD (0.012 mol-kg^{-1})	0.07	-0.04	-0.05	0.02	0.00	0.01
β-CD (0.012 mol-kg^{-1})	0.00	-0.04	-0.04	0.00	0.00	-0.02

* $\Delta\delta=\Delta\delta_{complexed} - \Delta\delta_{free}$.

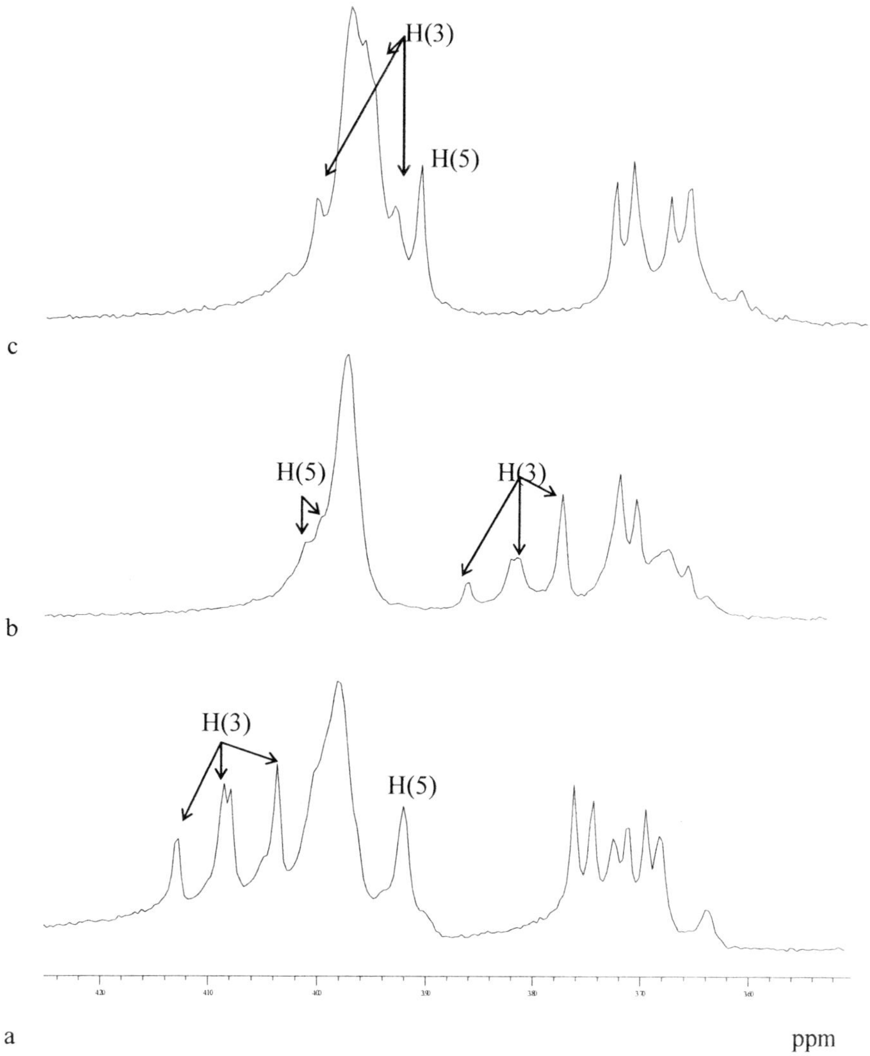

Figure 5. The partial ^{1}H NMR spectra of α-cyclodextrin (0.005 mol-kg^{-1}): a – in pure D_2O; b – in nicotinic acid solution (0.06 mol-kg^{-1}); c – in isonicotinic acid solution (0.06 mol-kg^{-1}) at 298.15 K and pH 3.4.

Figures 5 and 6 illustrate the partial ^{1}H NMR spectra of α- and β-cyclodextrin alone and in the presence of excess amounts of pyridinecarboxylic acids. For α-cyclodextrin, addition of the nicotinic acid induces the significant upfield shift of signal from H(3) proton ($\Delta\delta H_{(3)}$=-0.27 ppm) and less pronounced downfield shifting of signal from H(5) proton ($\Delta\delta_{H(5)}$=0.07 ppm). In the presence of the isonicotinic acid, the signals of H(3) and H(5) α-cyclodextrin protons exhibit upfield ($\Delta\delta_{H(3)}$=-0.12 ppm) and downfield ($\Delta\delta_{H(5)}$=0.02 ppm) shift, respectively. In case of both acids the negligible Δδ values (≈-0.01 ppm) for H(1), H(2), H(4) and H(6) protons of α-cyclodextrin were obtained. The less noticeable Δδ values observed for interaction of α-cyclodextrin with isonicotinic acid confirm occurring of weaker binding in this system Thus, the measurable shifting observed only for H(3) and H(5) protons confirms the inclusion of nicotinic and isonicotinic acids into α-cyclodextrin cavity. The shielding of

H(3) and deshielding of H(5) suggest that deep inclusion of pyridinecarboxylic acids takes place. Carboxylic group is completely included in α-cyclodextrin cavity and located closer to the narrow rim, whereas the aromatic ring lies in the plane of the six H(3) protons. The same binding mode was obtained by Simova and Schneider [27] for complexation of substituted benzoic acids and benzoate anions by α-cyclodextrin. As it was found, the deep penetration of the carboxylic group of substituted benzoic acids, irrespective of the substituent nature, takes place in the direction from the wider rim of α-cyclodextrin molecule. Thus, H(3) protons of α-cyclodextrin are located in the shielding area of aromatic ring, while the H(5) protons are deshielded by the carboxylic group.

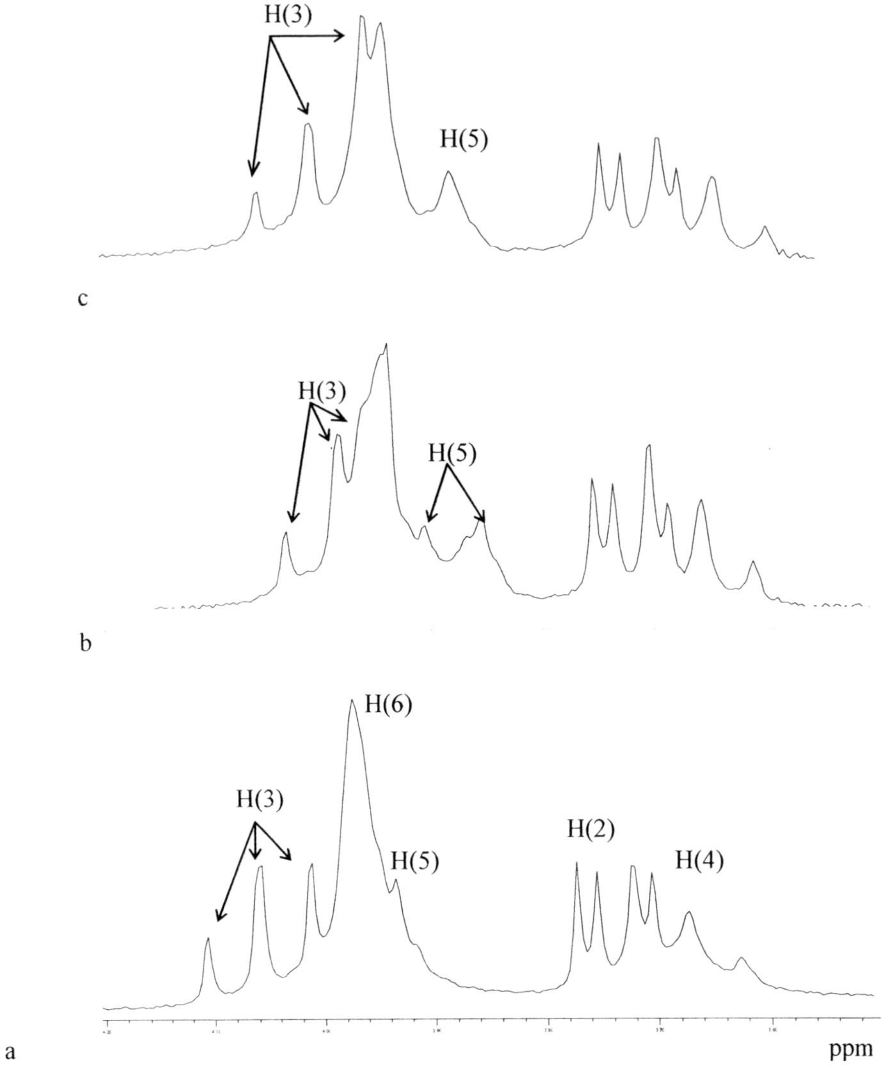

Figure 6. The partial ^{1}H NMR spectra of β-cyclodextrin (0.005 mol-kg^{-1}): a – in pure D_2O; b – in nicotinic acid solution (0.06 mol-kg^{-1}); c – in isonicotinic acid solution (0.06 mol-kg^{-1}) at 298.15 K and pH 3.4.

The much smaller $\Delta\delta$ were obtained for the interactions of β-cyclodextrin with nicotinic acid ($\Delta\delta_{H(3)}$=-0.07 and $\Delta\delta_{H(5)}$=-0.08 ppm) and isonicotinic acid ($\Delta\delta_{H(3)}$=-0.04 and $\Delta\delta_{H(5)}$=-0.04 ppm). The magnitude and sign of $\Delta\delta$ for H(3) and H(5) are practically the same. Therefore, these results do not unambiguously indicate the binding mode of β-cyclodextrin with pyridinecarboxylic acids under study. The most probable situation is formation of two kinds of complexes in which the aromatic ring is shallow inserted from both wide and narrow rims of β-cyclodextrin molecule.

Interactions of Cyclodextrins with Nicotinamide

Substitution of the carboxylic group of the nicotinic acid molecule on its amide influences the complexation process. As can be seen from concentration dependences of enthalpy of transfer presented in Figure 7 the weak binding of nicotinamide with α- and β-cyclodextrin takes place at pH 6.0 when the nicotinamide exist as uncharged molecule. The linear dependences obtained for interactions of cationic (pH=2.0) and neutral (pH=6.0) species of nicotinamide with hydroxypropyl-α-cyclodextrin points out the interactions without complex formation.

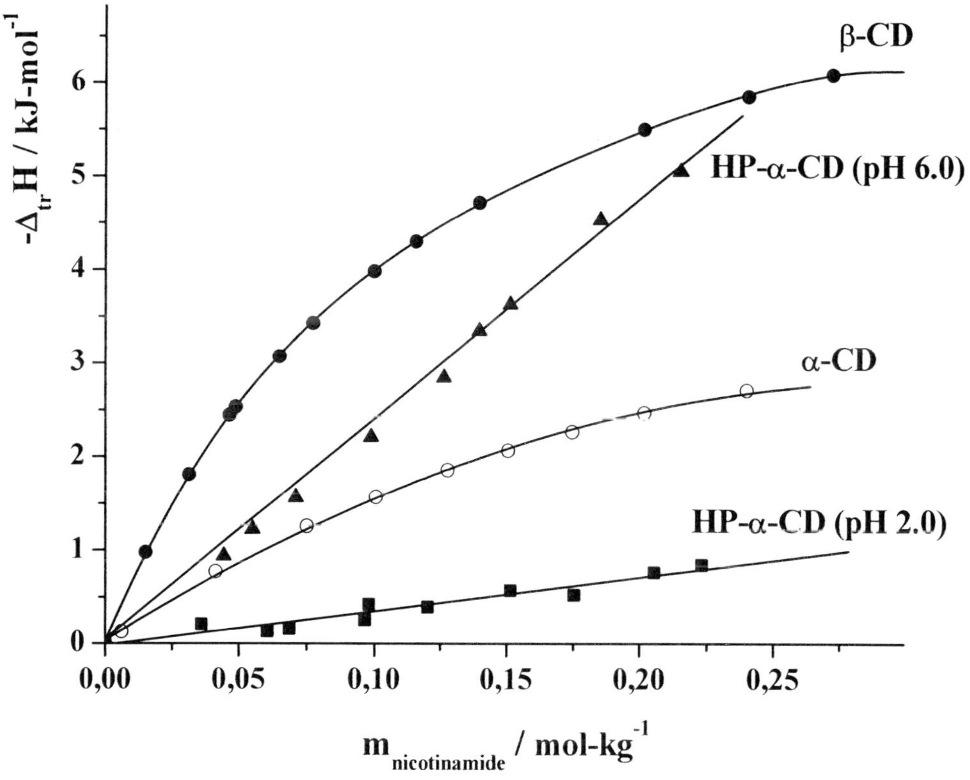

Figure 7. Enthalpy of transfer of cyclodextrins from water to aqueous solutions of nicotinamide *versus* the nicotinamide concentration (298.15 K).

Thermodynamic parameters of complex formation of nicotinamide with α- and β-cyclodextrin are listed in Table 2. Complexes are weak and stabilized only by the enthalpy term. Comparison of the thermodynamic parameters of complex formation of α-cyclodextrin with nicotinic acid and its amide allows to note the weaker binding affinity of α-cyclodextrin to nicotinamide. Enthalpy and entropy contributions in the free energy value are considerably lower in this case. The same tendency is characteristic for chemical shift changes obtained from ^{1}H NMR measurements (Figure 8) In particular, the small upfield shift ($\Delta\delta_{H(3)}$=-0.06 ppm) obtained only for H(3) protons of α-cyclodextrin (Figure 8) indicates the shallow insertion of nicotinamide into macrocylic cavity from the wider rim of host molecule. On the contrary, the deep inclusion of the nicotinic acid was detected.

The opposite behavior of β-cyclodextrin towards the nicotinic acid and nicotinamide was found. As it was discussed above, penetration of the nicotinic acid into β-cyclodextrin cavity occurs, however, the stable inclusion complex is not formed. In case of the nicotinamide, more exothermic interaction resulting in complex formation was observed. The signals form H(3) and H(5) β-cyclodextrin protons are upfield shifted in the ^{1}H NMR spectra shown in Figure 9. Practically the same magnitude of shifting of signals from these protons ($\Delta\delta_{H(3)}$=-0.07 and $\Delta\delta_{H(5)}$=-0.10 ppm) allows to suggest that nicotinamide can penetrate inside the β-cyclodextrin cavity from the wide and narrow rims.

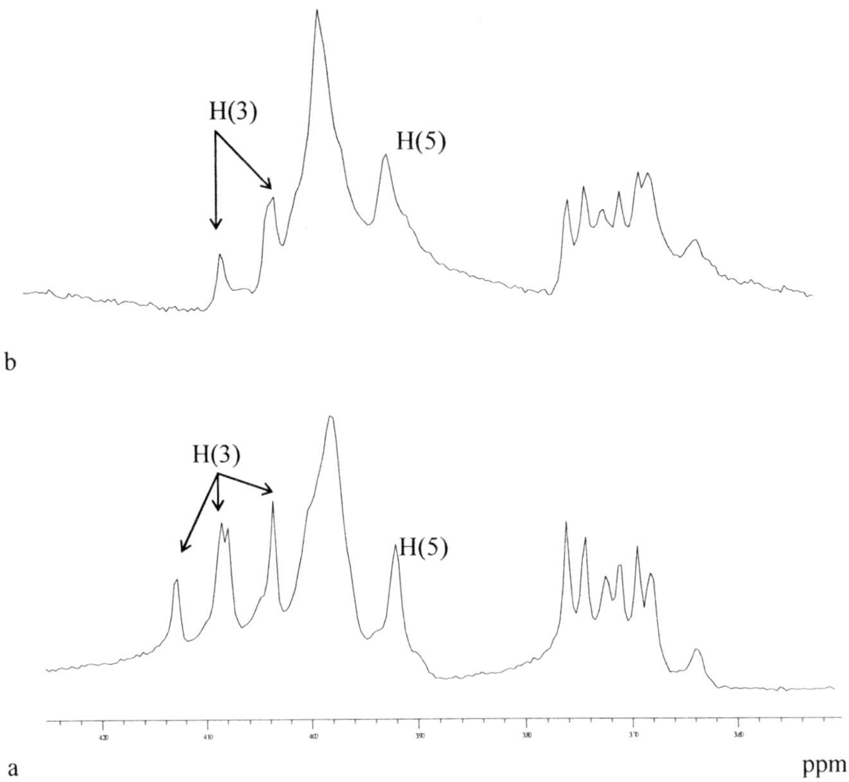

Figure 8. The partial ^{1}H NMR spectra of α-cyclodextrin (0.005 mol-kg^{-1}): a – in pure D_2O; b – in nicotinamide solution (0.06 mol-kg^{-1}) at 298.15 K.

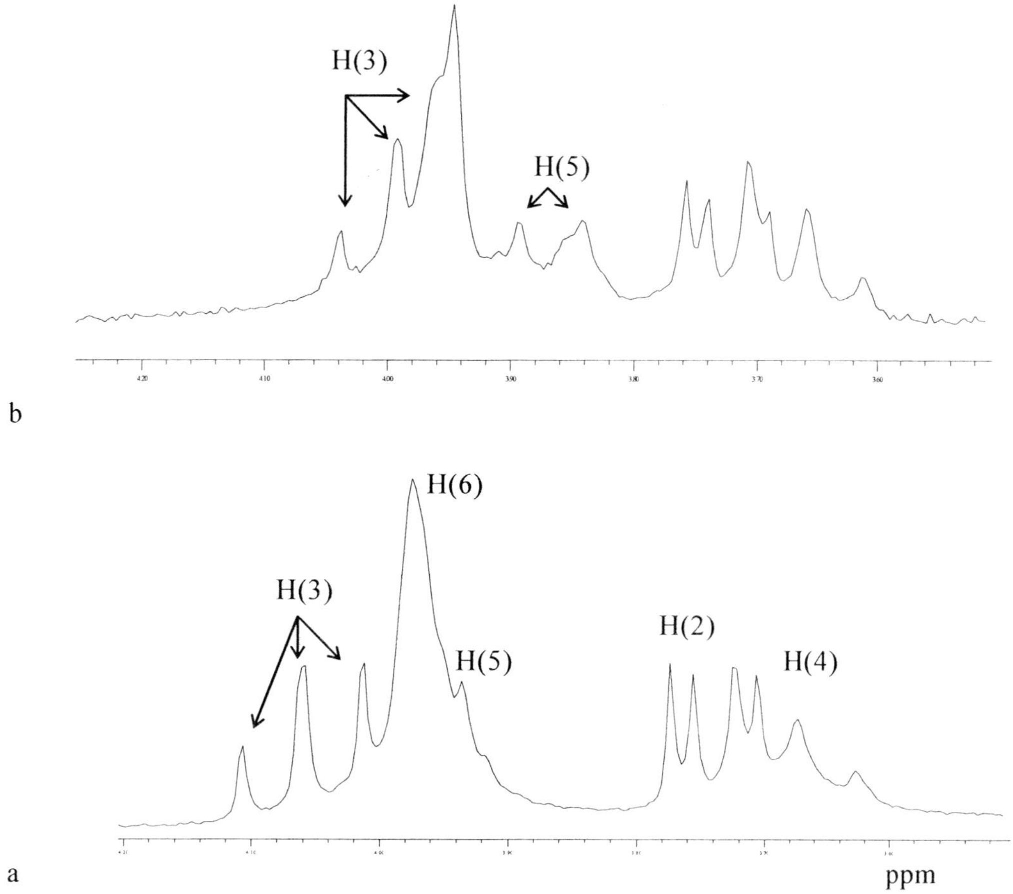

Figure 9. The partial ^{1}H NMR spectra of β-cyclodextrin (0.005 mol-kg^{-1}): a – in pure D_2O; b – in nicotinamide solution (0.06 mol-kg^{-1}) at 298.15 K.

The partial hydroxypropylation of OH-groups surrounding the α-cyclodextrin cavity prevents the process of complex formation with nicotinamide due to the possible steric hindrance arisen from these bulky substituents.

It should be also noted that protonation of nitrogen of the pyridine ring results in decreasing of exothermicity of interaction. The h_{xy} obtained for interaction of hydroxypropyl-α-cyclodextrin with nicotinamide at pH 2.0 and 6.0 differ (Table 1). The absence of binding of protonated nicotinamide with α-, hydroxypropyl-α- and β-cyclodextrin was also confirmed by the capillary electrophoresis (see Figure 10 as example). It can be caused by the stronger hydration of positively charged cationic form of the nicotinamide in comparison with its neutral form. It should be emphasized here that the similar influence of the nitrogen protonation on the thermodynamic parameters of complex formation of α-cyclodextrin with nicotinic acid was observed (Table 2).

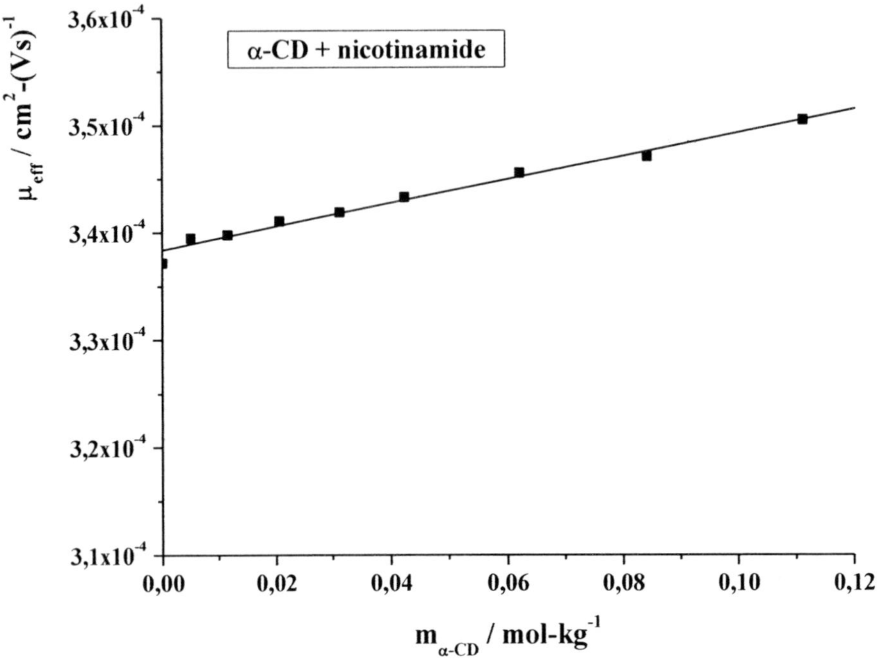

Figure 10. Dependences of viscosity-corrected effective mobility (μ_{eff}) of nicotinamide (1.5 mmol-kg^{-1}) and nicotinic acid (1.5 mmol-kg-1) from α-cyclodextrin concentration (phosphate buffer pH 2.0; T=298.15 K). Insert presents the system with complex formation and is shown for comparison.

Interactions of Cyclodextrins with Pyridoxine and Pyridoxal

Pyridoxine and pyridoxal having numerous polar substituents in the pyridine ring do not participate in the binding with cyclodextrins under study as it was determined by titration calorimetry at pH 6.8 and capillary electrophoresis at pH 2.0. Figures 11 and 12, as examples, illustrate the linear experimental dependences obtained by different methods. Figures 11, 12 show that the nitrogen protonation has no influence on the ability of pyridoxine and pyridoxal to form complexes with cyclodextrins. Complex formation of cyclodextrins with pyridoxine and pyridoxal does not occur due to the following possible reasons. First of them, is geometric discrepancy between cavity diameter and guest molecule dimensions. The numerous substituents in the pyridine ring may be too bulky hindering the inclusion of pyridoxine and pyridoxal into the cyclodextrin cavity. The second reason, is the high hydrophilicity and strong solvation of these guests.

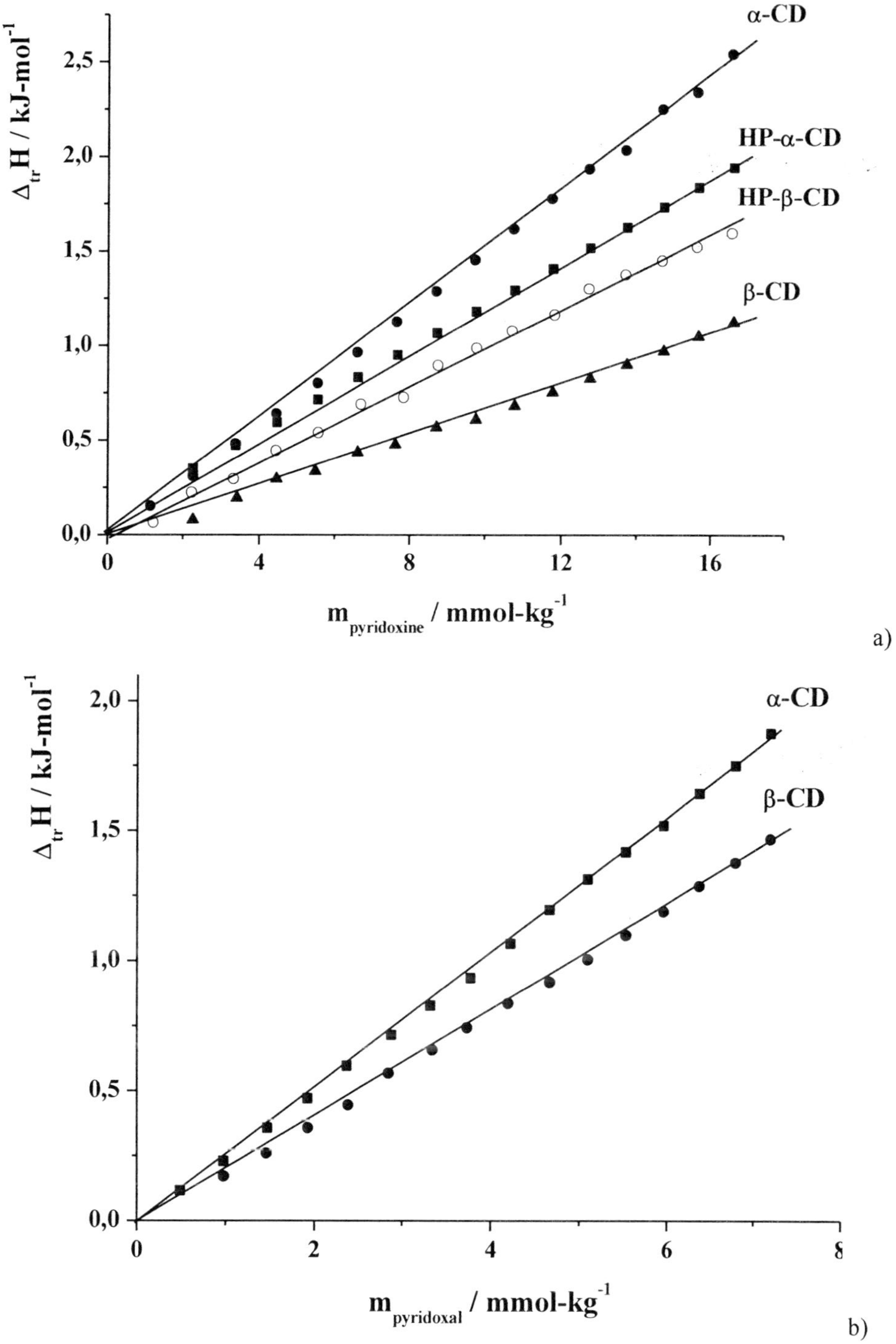

Figure 11. Enthalpy of transfer of cyclodextrins from water to vitamin B_6 solutions *versus* the vitamin B_6 concentration (298.15 K, pH 6.8): a- pyridoxine; b – pyridoxal.

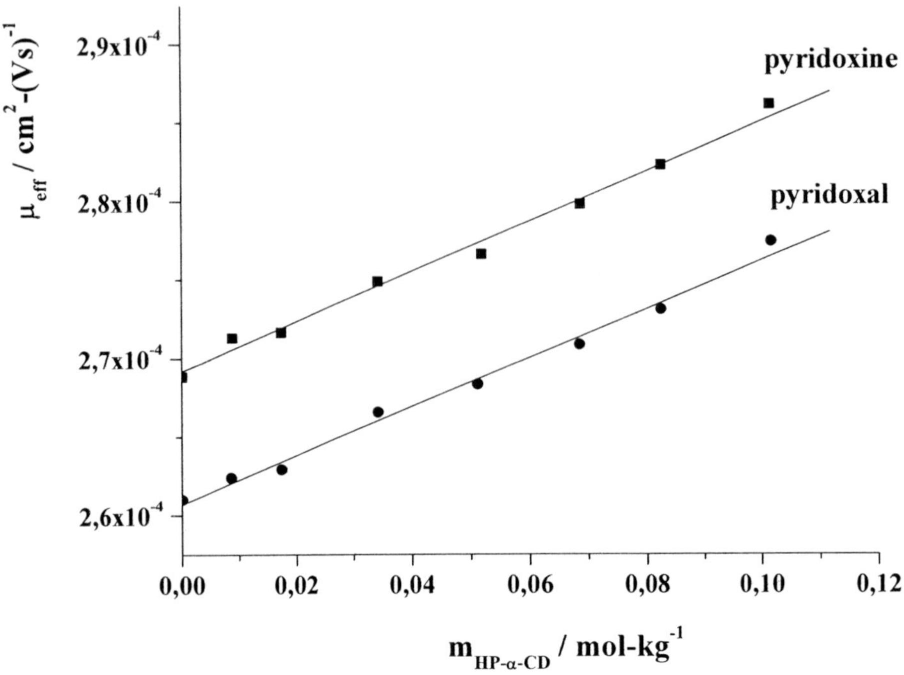

Figure 12. Dependences of viscosity-corrected effective mobility (μ_{eff}) of vitamins B_6 (1.5 mmol-kg^{-1}) from hydroxypropyl-α-cyclodextrin concentration (phosphate buffer pH 2.0; T=298.15 K).

Weak non-bonding interactions of pyridoxine and pyridoxal with cyclodextrins were characterized by the enthalpic coefficients, the positive values of which indicate the predominance of endothermic effects due to the dehydration of the solutes (Table 1). The enthalpic coefficients of pair interactions of β-cyclodextrin with pyridoxine and pyridoxal are considerably lower than the corresponding ones for systems with α-cyclodextrin. This fact can be explained by the best size-match of pyridoxine and pyridoxal to the β-cyclodextrin cavity. Thus, the larger cavity of β-cyclodextrin is more favorable for B_6 vitamins having numerous substitutes in structure, and attractive interactions characterized by exothermic effects and resulting in a decrease of h_{xy} play a noticeable role. Introduction of hydroxypropyl groups in the α- and β-cyclodextrin molecules does not promote the binding of these hosts with pyridoxine and pyridoxal.

CONCLUSION

The present study showed that α- and β-cyclodextrins display selectivity in the binding with B-vitamins in aqueous solutions at 298.15 K. Formation of the following inclusion complexes: α-cyclodextrin/nicotinic acid, hydroxypropyl-α-cyclodextrin/nicotinic acid, α-cyclodextrin/isonicotinic acid, α-cyclodextrin/nicotinamide and β-cyclodextrin/nicotinamide was detected. Processes of complex formation are enthalpy driven. Complexes are characterized by the relatively low stability constants the values of which lie in the range 3-33 kg-mol^{-1}.

The processes of complex formation are sensitive to the guest structure and cyclodextrin cavity dimensions. In particular, α-cyclodextrin forms more stable complexes with the nicotinic acid than with its structural isomer (isonicotinic acid) and its amide (nicotinamide). Ionization of the carboxylic group of the pyridinecarboxylic acids under study governs their binding with cyclodextrins, whereas the protonation of the nitrogen of the pyridine ring has no significant influence on the complex formation process. The presence of more than one polar side groups in the pyridine ring results in enthalpically unfavorable interactions of pyridoxine and pyridoxal with cyclodextrins.

The cyclodextrin cavity diameter plays an important role in complex formation with B-vitamins. The deep insertion of nicotinic and isonicotinic acids into α-cyclodextrin cavity takes place, while only shallow penetration inside the β-cyclodetrin cavity is observed. To the contrary, increasing of the macrocyclic cavity size promotes the more enthalpically favorable interaction of β-cyclodextrin with nicotinamide and formation of more stable inclusion complex in this system. Introduction of hydroxypropyl substituents into cyclodextrin molecule weakens the binding and makes it less exothermic. Probably, the bulky hydroxypropyl groups are more solvated and obstruct the penetration of a guest into the cyclodextrin cavity.

ACKNOWLEDGMENT

Author thanks Prof. W. Zielenkiewicz (Institute of Physical Chemistry of Polish Academy of Sciences, Warsaw, Poland), Prof. S. Milioto (Department of Physical Chemistry, University of Palermo, Italy) and Prof. G.K.E. Scriba (Department of Pharmaceutical Chemistry, University of Jena, Germany) for fruitful collaboration. This work was financially supported by DAAD, Russian Science Support Foundation and Russian Foundation for Basic Research (grant no. 03-03-96411).

REFERENCES

[1] Szejtli, J. Cyclodextrins and their inclusion complexes. Budapest: Akadémiai Kiadó; 1982.

[2] Bergeron, RJ. Cycloamylose-substrate binding. In: Atwood JL, Davies JED, MacNicol DD editors. Inclusion compounds. London: Academic Press; 1984; 391-443.

[3] Connors, KA. The stability of cyclodextrin complexes in solution. *Chem. Rev.* (1997) 97 1325-1357.

[4] Liu, L; Guo, Q-X. The driving forces in the inclusion complexation of cyclodextrins. *J. Inclus. Phenom. Macrocycl. Chem.* (2002) 42, 1-14.

[5] Khan, AR; Forgo, P; Stine, KJ; D'Souza, VT. Methods for selective modifications of cyclodextrins. *Chem. Rev.* (1998) 98, 1977-1996.

[6] Gattuso, G; Nepogodiev, SA; Stoddart, JF. Synthetic cyclic oligosaccharides. *Chem. Rev.* (1998) 98, 1919-1958.

[7] Hedges, AR. Industrial applications of cyclodextrins. *Chem. Rev.* (1998) 98, 2035-2044.

[8] Uekama, K; Hirayama, F; Irie, T. Cyclodextrin drug carrier systems. *Chem. Rev.* (1998) 98, 2045-2076.

[9] Martin Del Valle, EM. Cyclodextrins and their uses: a review. *Process Biochemistry* (2004) 39, 1033-1046.

[10] Szejtli, J. Industrial applications of cyclodextrins. In: Atwood JL, Davies JED, MacNicol DD editors. Inclusion compounds. London: Academic Press; 1984; 331-390.

[11] Chankvetadze, B. Capillary electrophoresis in chiral analysis. John Wiley@sons Ltd.; 1997; 555 p.

[12] Vencina, MC; Fermier AM; Guo Y. Comparing cyclodextrin derivatives as chiral selectors for enantiomeric separation in capillary electrophoresis. *J. Chrom. A* (2002) 973, 187-196.

[13] Brown, MS; Goldstein, JL. Drugs used in the treatment of hyperlipoproteinemias. In: Gilman AG, Rall T, Nies AS, Taylor P editors. Goodman and Gilman's the pharmacological basis of therapeutics. New York: Pergamon Press Inc.; 1990; 874-896.

[14] Gilman, AG; Goodman, LS; Gilman, A. Pharmacological basis of therapeutics. MacMillan. New York; 1985.

[15] Lewis, EA; Hansen, LD. Thermodynamics of binding of guest molecules to α- and β-cyclodextrins. *J. Chem. Soc. Perkin Trans. 2* (1973) 2081-2085.

[16] Gezawi, SE; Omar, N; Rabbat, NE; Ueda, H; Perrin, JH. Microcalorimetric and chromatographic investigations of the binding of some pyridine derivatives to cyclodextrins. *J. Pharm. Biomed. Anal.* (1988) 6, 399-406.

[17] Fornasier, R; Marcuzzi, F; Zorzi, D. High selectivity in the catalytic hydrogenation of acyl-substituted pyridines in alkaline solution: effect of complexation by β-cyclodextrin. *J. Molec. Catal.* (1987) 43, 21-26.

[18] Ferčej-Temelyotov, D; Kmet, M; Kocjan, D; Kotnik, S; Resman, A; Urleb, U; Verhnjak, K; Zver, I; Źmitek, J. Stereoselectivity in β-cyclodextrin complexation of 1,4-dihydropyridine derivatives. *Chirality* (1993) 5, 288-292.

[19] Terekhova, IV; Obukhova, NA. Thermodynamics of nicotinic acid interactions with some saccharides. *J. Solution Chem.* (2005) 34, 1273-1282.

[20] Zielenkiewicz, W; Terekhova, IV; Marcinowicz, A; Koźbial, M; Poznanski, J. Interactions of native and modified cyclodextrins with some B-vitamins. Calorimetric and ^{1}H NMR study. *J. Therm. Anal. Calorim.* (2007) doi: 10.1007/s10973-006-7974-7.

[21] Terekhova, IV; Obukhova, NA; Kumeev, RS; Al'per, GA. Interactions of α- and hydroxypropyl-α-cyclodextrins with nicotinic acid in aqueous solutions. *Russ. J. Phys. Chem.* (2005) 79, 1976-1979.

[22] Terekhova, IV; Scriba, GKE. Study on complex formation of biologically active pyridine derivatives with cyclodextrins by capillary electrophoresis. *J. Pharm. Biomed. Anal.* (2007) 45, 688-696.

[23] Terekhova, IV; Kulikov, OV; Titova, ES. Enthalpic characteristics of interactions occurring between an ascorbic acid and some saccharides in aqueous solutions. *Thermochim. Acta* (2004) 412, 121-124.

[24] McMillan, WG; Mayer, JE. The statistical thermodynamics of multicomponent systems. *J. Chem. Phys.* (1945) 13, 276-305.

[25] Desnoyers, JE; Perron, G; Avédikian, L; Morel, J-P. Enthalpies of the urea-*tret*-butanol-water system at 25 ^{0}C. *J. Solut. Chem.* (1976) 5, 631-644.

[26] Koczoń, P; Dobrowolski, JCz; Lewandowski, W; Mazurek, AP. Experimental and theoretical IR and Raman spectra of picolinic, nicotinic and isonicotinic acids. *J. Mol. Struct.* (2003) 655, 89-95.

[27] Simova, S; Schneider, H-J. NMR analyses of cyclodextrin complexes with substituted benzoic acids and benzoate anions. *J. Chem. Soc., Perkin Trans. 2* (2000) 1717-1722.

[28] Salvatierra, D; Jaime, C; Virgile, A; Sanchez-Ferrando, F. Determination of the inclusion geometry for the β-cyclodextrin/benzoic acid complex by NMR and molecular modeling. *J. Org. Chem.* (1996) 61, 9578-9581.

[29] Stalin, T; Rajendiran, N. Intramolecular charge transfer effects on 3-aminobenzoic acid. *Chem. Phys.* (2006) 322, 311-322.

[30] Stalin, T; Shanthi, B; Vasantha Rani, P; Rajendiran, N. Solvatochroism, prototropism and complexation of para-aminobenzoic acid. *J. Incl.. Phenom. Macrocycl. Chem.* (2006) 55, 21-29.

[31] Schneider, H-J; Hacket, F; Rüdiger, V; Ikeda, H. NMR studies of cyclodextrins and cyclodextrin complexes. *Chem. Rev.* (1998) 98, 1755-1785.

In: Solution Chemistry Research Progress
Editor: Darian V. Bostrelli, pp. 115-131
ISBN 978-1-60021-993-1

Chapter 5

DEPENDENCE OF THE GIBBS ENERGY OF ADSORPTION OF NEUTRAL ORGANOPHOSPHOROUS COMPOUNDS ON EMPIRICAL INDICES OF BINARY SOLVENT MIXTURES

A. G. Anastopoulos
Department of Chemistry, Aristotle University of Thessaloniki
54124 Thessaloniki, Greece

ABSTRACT

Empirical indices of binary aqueous mixtures of protic and aprotic solvents, representing the electron pair and hydrogen bond exchange ability of these mixtures are correlated to the standard Gibbs energy of adsorption, ΔG^{o}_{ads}, of neutral organophosphorous compounds at the dropping mercury electrode - solution interface.

Correlations are carried out in terms of linear and multilinear empirical relations employing the Kamlet – Taft β_{KT}, α and the Dimroth – Reichardt $E_T(30)$ and Kosower Z parameters, which account for solvent – electrode and solvent – solvent interactions. On the basis of the heterosolvation assumption the effect of each binary mixture on the adsorption equilibrium, expressed by the ΔG^{o}_{ads}, is discussed. It is suggested that the effect of solvent composition on the values of ΔG^{o}_{ads} is predominately determined by the adsorption of organic cosolvent and the water-organic solvent interactions.

INTRODUCTION

The study of solvent effects on various electrochemical processes occurring in binary aqueous and purely non aqueous solvent mixtures is a matter of particular interest in many fields of science and advanced technology. One of the significant advantages of such studies is that solvent effects can be controlled by continuously changing the properties of the solvent mixture as the result of simply changing its composition.

A good deal of the work done in this research field aims to the connection between processes occurring at the electrode/solution interface and the properties of bulk mixtures. This task is usually a difficult one because of the selective solvation of ions and molecules and also because of the different solvating environment existing at the interface and the bulk, due to the stronger specific adsorption of the one, usually the organic, of the two cosolvents.

In the case of electrosorption phenomena almost all the research interest has, up to now, been focused to the correlation of the adsorption of non hydrophobic anions at the electrode/solution interface with processes and properties relating to the solvation behaviour of the solvent system.

As the first significant contribution to the study of these issues we must refer to the works of B.E.Conway [1-4] who connected the electronic interaction effects between metal electrodes and lone electron pairs with the necessary changes of the geometry of the inner solvation shell of ions during the specific adsorption of non hydrophobic anions.

According to Conway, in order that a symmetrically solvated anion to be specifically adsorbed on a metal surface, it is more likely not to reject a number of solvating molecules but to undergo a deformation of its solvation geometry, to allow for one or more of its lone pair orbitals to overlap with the emerging or vacant orbitals of the electrode surface. In general, the greater the size ratio of anions with respect to solvent molecules, the easier the deformation of the solvation structure. On the contrary the complete rejection of a solvent molecule from the inner solvation shell is energetically impractical.

A substantially different approach to the study of the effect of solvent properties on the ionic adsorption processes at the electrode-solution interface consists in using the various empirical parameters which usually express the polarity and the hydrogen bond and electron pair exchange ability of the solvent system.

Fawcett and coworkers [5] correlate the Gibbs energy of adsorption of Br^- ions with parameters expressing the donor-acceptor properties of the solvent. The adsorption of anions A^- involves the replacement of solvent molecules from the interfacial layer according to the reaction:

$$A_b^- + S_{ads} \rightleftarrows A_{ads}^- + S_b \tag{1}$$

where S represents the solvent and the subscripts ads and b the states of adsorption and bulk solution respectively.

Therefore the standard Gibbs energy of adsorption may be written as [6,7]:

$$\Delta G_{ads}^{o} = G^{o}(A^{-})_{ads} + G^{o}(S)_{b} - G^{o}(S)_{ads} - G^{o}(A^{-})_{b} \tag{2}$$

where the quantities of the type $G^o(i)_j$ represent the standard Gibbs energy of species i at the position j. Among the quantities of the right hand side of equation (2) only the energy $G^o(A^-)_{ads}$, representing the ion-metal electrode interactions, does not practically depend on the change of the solvent composition. All the other energetic quantities of equation (2) are solvent depended, viz. the quantity $G^o(A^-)_b$ which represents the contribution of the changes of solvation due to the change of solvent and also the difference $G^o(S)_{ads} - G^o(S)_b$ representing the change of Gibbs energy produced by the adsorption of the solvent itself. On the basis of these assumptions ΔG^o_{ads} is correlated with parameters expressing the Lewis behavior of the solvent as for example the solvent acidity parameter $E_T(30)$ of Dimroth and Reichardt [8,9] and the donor number DN which is a measure of Lewis basicity. The above correlations are physically justified by the finding that the ΔG^o_{ads} of Br^- ions is increased by the decrease of solvation energy, expressed by the parameter $E_T(30)$, and by the weakening of the interactions between the electrode and the solvent molecules expressed by the DN.

Correlations of similar character and comparisons of interfacial parametrs of Cl^-, Br^-, I^- and SCN^- ions at the interface between non aqueous solvents and various metal electrodes, viz. Hg, Bi, Ga, In-Ga and Tl-Ga had been the subject of several works of Damaskin and coworkers [6,10]. In terms of the following equation (3) which is similar to (1):

$$nS_{ads} + A_b^- \rightleftarrows A_{ads}^- + nS_b \tag{3}$$

the Gibbs energy of adsorption of A^- ions is given by the relation:

$$\Delta G_{ads} = \Delta G_{M\text{-}A^-} - n\Delta G_{M\text{-}S} - \Delta G_{A^-\text{-}S} + n\,\Delta G_{S\text{-}S} \tag{4}$$

which is again substantially similar to (2). It is assumed that the quantities $\Delta G_{M\text{-}S}$ and $\Delta G_{S\text{-}S}$ do not depend on the nature of the adsorbed ion. It is also found out that the change of ΔG_{ads} between different solvents is mainly determined by the difference in ion solvation energy, while the contribution of $\Delta G_{M\text{-}S}$ is usually not large.

However, though the correlation of interfacial properties of adsorbed anions with solvent properties has been the subject of systematic research, the same is not true for the electrosorption of neutral molecules and large organic cations from pure and mixed solvents. Neutral and cationic organophosphorous and organoarsenic compounds are often markedly adsorbed from pure non aqueous solvents and binary aqueous mixtures. Such equilibrium adsorption data [11-19] can be used for interesting correlations between interfacial and bulk properties of adsorbate - solvent systems. In previous papers [11-12] we have examined the correlation of standard Gibbs energy of adsorption of various alkyl and phenyl substituted phosphine oxides at the mercury electrode with empirical solution parameters primarily in pure and secondarily in mixed solvents.

In the present paper our attention is drawn to binary aqueous-organic mixtures, while the adsorbates used are triphenyl and alkyl-diphenylphosphine oxides which are fairly soluble in the solvent systems studied. Our selection of the binary aqueous-organic systems was guided by the availability of bibliographic data [20-23] for the properties of these solvent systems and also by the possibility to use organic cosolvents which are strongly adsorbed at the Hg surface from mixed aqueous solutions but have a different effect on the water structure and the hydrogen bond and electron pair exchange ability of the mixtures. The strong adsorbability of organic cosolvents is related to the heterosolvation assumption which means that, independently of the composition of the bulk mixture, the interface is occupied by adsorbed organic solvent molecules within the potential range where the phosphine oxides are also adsorbed.

In solutions of neat solvents the adsorption reaction of the neutral adsorbate A and the corresponding equation for ΔG^{o}_{ads} look like the expressions (1) και (2) respectively for the adsorption of anions. However, in this case the quantity $G^{o}(A)_{ads} - G^{o}(A)_{b}$ may be considered as practically independent from the solvent because the adsorbate - metal electrode interaction energy is very weakly dependent on the change of the solvent and the solvation energy of the neutral adsorbate is generally low and practically independent of the solvent composition. Therefore it seems that in the case of non ionic adsorption in pure non aqueous solvents the change of the solvent mainly affects the quantity $G^{o}(S)_{ads} - G^{o}(S)_{b}$ which represents the change of the Gibbs energy associated with solvent adsorption.

In the case of the mixtures of two solvents S_1 and S_2 equation (2) must be modified in order to account for the interactions in which the two solvents are involved. Thus the description of the adsorption processes is getting more complex. Starting from a suggestion of Galus and coworkers [24] for electrode processes in mixed solvents involving reactant adsorption and replacing reactant by the neutral adsorbate, the solvation of which both at the interface and the bulk is assumed to be negligible, the adsorption reaction may be written as follows:

$$A_b + aS_{1ads} + bS_{2ads} \rightleftarrows A_{ads} + cS_{1b} + dS_{2b} \quad (5)$$

Where the subscripts ads and b denote the interface and the bulk respectively, while the numbers of the molecules of solvents S_1 and S_2, which initially were at the interface and then return to the bulk solution upon A adsorption, are correspondingly represented by the stoichiometric coefficients a,b και c,d. According to (5) the overall equilibrium is accompanied by a change of the standard Gibbs energy of adsorption:

$$\Delta G^{o}_{ads} = \Delta G^{o}_{ads}(A) - \Delta G^{o}_{ads}(S_1) - \Delta G^{o}_{ads}(S_2) \quad (6)$$

In equation (6) the meaning of the right side terms is the following. Change of the overall Gibbs energy of adsorption due to the adsorption of species A:

$$\Delta G^{o}_{ads}(A) = G^{o}(A)_{ads} - G^{o}(A)_{b} \quad (7)$$

Change of the overall Gibbs energy of adsorption due to the adsorption of solvent S_1:

$$\Delta G^o_{ads}(S_1) = G^o(S_1)_{ads} - G^o(S_1)_b \tag{8}$$

Change of the overall Gibbs energy of adsorption due to the adsorption of solvent S_2:

$$\Delta G^o_{ads}(S_2) = G^o(S_2)_{ads} - G^o(S_2)_b \tag{9}$$

On the basis of the heterosolvation assumption, at the potential range of maximum adsorption of the studied phosphine oxides the adsorption of the organic cosolvent, S_2, predominates over water, S_1. Thus, only the term $\Delta G^o_{ads}(S_2)$ has a remarkable contribution to the overall change of ΔG^o_{ads} during the change of the solvent mixture composition. On the other hand, the term $\Delta G^o_{ads}(A)$ does not markedly depend on the mixture composition. Moreover, the contribution of water adsorption, expressed by the term $\Delta G^o_{ads}(S_1)$ is small and approximately independent of the mixture composition within the potential range where the phosphine oxide and S_2 adsorption prevails. Therefore, equation (6) is satisfactorily approximated by relation (10):

$$\Delta G^o_{ads} \approx G^o(S_2)_b - G^o_{ads}(S_2)_S \tag{10}$$

where the term $G^o(S_2)_{ads}$ represents the organic solvent - electrode interactions, which may be considered to relate with the ability of individual water molecules to act as Lewis base to the Hg atoms. The Lewis basicity is usually measured by the donor number DN [9,25-27]. However, very few data for the DN of solvent mixtures are found in the literature. The Kamlet - Taft parameter β_{KT}, which is a measure of the electron pair donor ability of the mixture, may be used as a substitute of DN. The term $G^o(S_2)_b$ of expression (10) generally represents the bulk interactions not only of the type S_2-S_2 but also of the type S_1-S_2. Therefore it may be written as:

$$G^o(S_2)_b = G^o(S_2 - S_2)_b + G^o(S_2 - S_1)_b \tag{11}$$

These interactions are mainly hydrogen bond exchange interactions and they are expressed by means of the parameter $E_T(30)$ of Dimroth - Reichardt [9], which primarily describes the ability of a solvent to act as a hydrogen bond donor and secondarily expresses the polarity of the solvent. Similarly to the $E_T(30)$ parameter, the Kosower's [29] parameter Z is a sensitive measure of the hydrogen bonding ability also incorporating contributions from the polarity and the electron pair donicity of the solvent. Free of polarity and electron pair donicity contributions is the Kamlet - Taft parameter α, which is a measure of the hydrogen bond donating ability (Lewis acidity).

Experimental Part

The standard Gibbs energies of adsorption of triphenylphosphine oxide (TPO), methyldiphenyphosphine oxide (MeDPO) and ethyldiphenylphosphine oxide (EtDPO) were calculated by means of isotherm analysis of double layer capacitance vs electrode potential data at the potential of maximum adsorption.

Double layer capacitance vs potential measurements were carried out by means of an arrangement comprising a DSP Lock-in amplifier (model SR830 from STANFORD RESEARCH) and POS73 Potentioscan Wenking potentiostat (from BANK ELEKTRONIK) interfaced by a GPIB connection to a Pentium PC running LabVIEW under Windows XP. The amplitude and the frequency of the ac signal were set to 4mV and 80Hz respectively. The measured capacitance values are reproducible within 1-1.5% and their mean % deviation from standard measurements of Grahame [30] was always better than 2%.

A 3-electrode electrochemical cell was used including a Pt foil auxiliary electrode, a dropping mercury electrode (DME) as working electrode with a flow rate of 1.04mg/s giving a free drop time of 8s and an Ingold 363-S7/120 Ag/AgCl/ 0.1M aq. NaCl reference electrode. The latter is fairly reversible in all mixtures up to 90% v/v in organic cosolvent. The measured potentials are reproducible by about 1%. All measurements were carried out at $25.0\pm0.5^{o}C$.

Base electrolyte $LiClO_4$ (Ventron GMBH), Methanol (MeOH; Fluka, puriss p.a.), Acetonitrile (AN; Merck, pro analysi) and Ethanol (Riedel-de Haën, puriss.pa) were used without further purification. Acetone (AC), water and Hg were used after double distillation. TPO (Merck; for synthesis) was purified by repeated recrystallization from 1:1 methanol-water mixture and then dried at 120°C. MeDPO (Sigma-Aldrich, 98%) was repeatedly recrystallized from 1:1 benzol-hexane mixture and then dried at 80°C. EtDPO (Sigma-Aldrich, 97%) was purified as quoted in [31].

Results

The values of the Gibbs energy of adsorption, ΔG^{o}_{max}, calculated at the potential of maximum adsorption of TPO, MeDPO and EtDPO in aqueous mixtures of acetone, acetonitrile, ethanol and methanol are shown in tables 1 to 4. All experimental values of ΔG^{o}_{max} are seen to decrease with increasing organic cosolvent content. This means that the ΔG^{o}_{max} of phosphine oxides decreases towards the direction of increasing adsorption of organic cosolvents . This direction maybe also described as the enhancement of the heterosolvation effect. The same result is also contributed by the increase of solubility of phospine oxides due to the increase of the organic solvent content.

In figures 1, 2 and 3 it is worth to examine the dependence of ΔG^{o}_{max} of MeDPO, EtDPO and TPO respectively from the composition of the solvent mixtures, which on a common basis is expressed by the molar fraction X_{org} , of the organic cosolvent in each system.

Among the studied phosphine oxides, over all the range of X_{org} values, the stronger adsorption i.e. the highest values of ΔG^{o}_{max} correspond to methanol and the lower ones to acetone. In general the adsorbability of phosphine oxides in the binary systems studied follows the order MeOH-H_2O>EtOH-H_2O> AC-H_2O.

The adsorbability of the phosphine oxides in the system AN-H_2O for $X_{AN}<0.20$ is somewhat similar to that in MeOH-H_2O mixtures, while for $X_{AN}>0.70$ it approximates the AC-H_2O system. The difference in the ΔG^{o}_{max} values from one system to another may be mainly attributed to the adsorption competition between the adsorbate and organic cosolvent taking in to account that the adsorbabilty series [32-35] of the studied solvents from mixed with water solutions is AC>AN>EtOH>MeOH. Within the same binary system the decrease of ΔG^{o}_{max} values by increasing the content of organic solvent is related to the predominance of the adsorption of organic solvent over the adsorbate. The latter effect may be understood in terms of the stronger electron pair donation ability of the organic solvent to the Hg electrode than the phosphine oxides and also, as previously mentioned, by the decrease of the hydrophobic interactions of phosphine oxides with increasing organic solvent content.

Table 1. Standard Gibbs energy of maximum adsorption of MeDPO, EtDPO and TPO in acetone- water mixtures and related solvent parameters. ΔG^{o}_{max} in kJ/mol

% AC	X_{AC}	$-\Delta G^{o}_{max}$ MeDPO	$-\Delta G^{o}_{max}$ EtDPO	$-\Delta G^{o}_{max}$ TPO	$E_T(30)$ kJ/mol	β_{KT}	Z	α
0	0.000	36.6	38.4	40.4	264.02	0.470	92.5	1.17
10	0.0268	35.1	35.9	36.9	258.71	0.51	91.9	1.131
20	0.0583	32.3	33.2	34.7	251.35	0.554	91.3	1.087
30	0.104	29.0	29.8	30.8	241.14	0.615	90.4	1.024
40	0.145	26.8	27.5	28.6	235.57	0.639	89.3	0.973
50	0.210	23.2	24.3	24.8	229.34	0.671	87.3	0.897
60	0.27	21.3	22.0	22.3	225.45	0.677	85.7	0.855
70	0.39	18.6	19.7	20.8	220.93	0.671	83.3	0.77
80	0.495	15.6	16.2	17.1	219.25	0.651	81.2	0.729
90	0.682	12.9	13.4	14.3	211.30	0.631	77.2	0.619
100	1	10.5	11.8	13.6	176.57	0.56	65.7	0.08

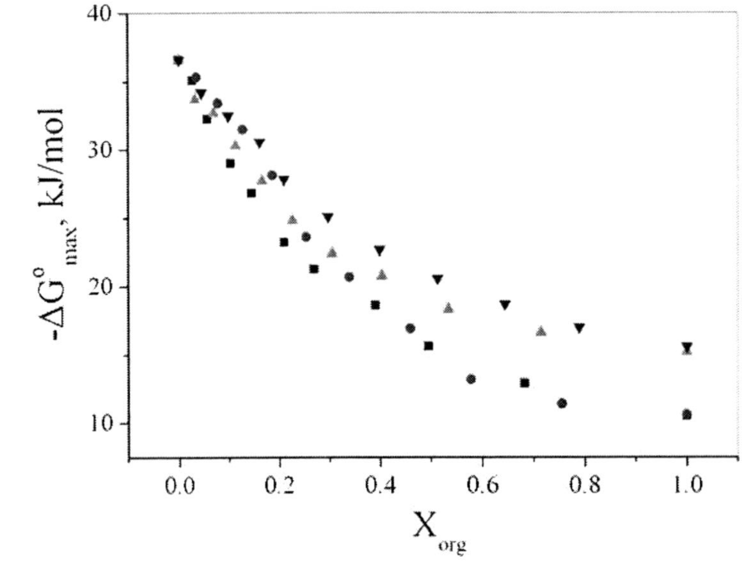

Figure 1. Scatter plots of ΔG^{o}_{max} of MeDPO against the molar fraction of organic cosolvent in 0.1M $LiClO_4$ binary solutions of water with: Acetone ■, Acetonitrile ●, Ethanol ▲ and Methanol ▼.

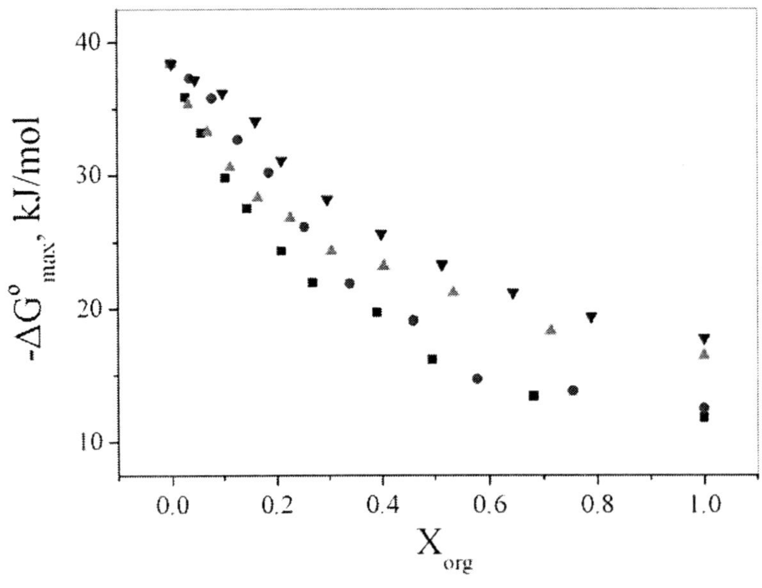

Figure 2. Scatter plots of ΔG^{o}_{max} of EtDPO against the molar fraction of organic cosolvent in 0.1M $LiClO_4$ binary solutions of water with: Acetone ■, Acetonitrile ●, Ethanol ▲ and Methanol ▼.

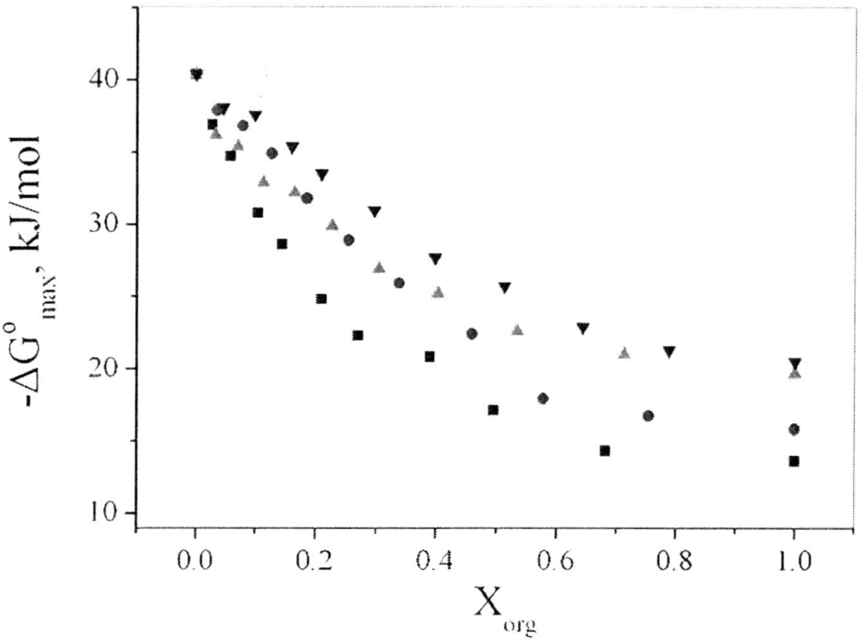

Figure 3. Scatter plots of ΔG^{o}_{max} of TPO against the molar fraction of organic cosolvent in 0.1M $LiClO_4$ binary solutions of water with: Acetone ■, Acetonitrile ●, Ethanol ▲ and Methanol ▼.

In general, the series AC>AN>EtOH>MeOH is also in agreement with the observed effect of the change of the mixture composition on ΔG^{o}_{max} values. At X_{org} values up to about 0.50 the greater slope of the point diagrams of figures 1,2,3 is observed in AC-H_2O mixtures and the smaller slope in MeOH-H_2O ones. At X_{AC}>0.50, due to the saturation of electrode surface, the composition dependence of ΔG^{o}_{max} is markedly weakened, while the corresponding weakening in methanol is clearly reduced. From the other two organic solvents ethanol, as more strongly adsorbed than methanol, presents, as expected, a stronger effect (greater slope) and lower ΔG^{o}_{max} phosphine oxide values from methanol only at low X_{org}. On the other side, it tends to approach methanol for X_{org}>0.50.

In mixtures with water the behaviour of acetonitrile, the adsorbability of which resembles that of acetone [35], is noticeably different from that of the other systems. At low contents, acetonitrile behaves similarly to methanol, while at high X_{AN} it approximates acetone [35]. This profile of behaviour may be related to the change of the internal structure of the system AN-H_2O, as long as AN similarly to AC is known to be a water structure beaker [36]. In general, for the mixtures AN+H_2O it has been suggested [37-39] that the increase of X_{AN} up to the value of 0.80 results to the creation of AN+H_2O clusters. At X_{org}>0.80 the structure of water is completely destroyed and the mixture tends to the structure of pure acetonitrile, which may be also true for acetone. Therefore, at high AN and AC contents the values of ΔG^o_{max} for the three phosphine oxides studied tend to coincide.

Table 2. Standard Gibbs energy of maximum adsorption of MeDPO, EtDPO and TPO in acetonitrile - water mixtures and related solvent parameters. ΔG^o_{max} in kJ/mol

%AN	X_{AN}	$-\Delta G^o_{max}$ MeDPO	$-\Delta G^o_{max}$ EtDPO	$-\Delta G^o_{max}$ TPO	$E_T(30)$ kJ/mol	β_{KT}	Z	α
0	0	36.6	38.4	40.4	264.02	0.47	92.5	1.17
10	0.036	35.3	37.3	37.9	258	0.553	91.7	1.127
20	0.079	33.4	35.8	36.8	252.89	0.584	91.1	1.078
30	0.128	31.5	32.7	34.9	247.66	0.611	90.2	1.028
40	0.186	28.1	30.2	31.8	242.39	0.611	89.5	0.976
50	0.255	23.6	26.1	28.9	237.75	0.610	88.4	0.94
60	0.339	20.7	21.9	25.9	235.07	0.606	87.4	0.908
70	0.46	16.9	19.1	22.4	232.98	0.594	86.7	0.872
80	0.578	13.2	14.7	17.9	231.89	0.593	85.5	0.834
90	0.755	11.4	13.8	16.7	226.53	0.573	80.9	0.738
100	1	10.6	12.5	15.8	192.47	0.47	71.1	0.25

Table 3. Standard Gibbs energy of maximum adsorption of MeDPO, EtDPO and TPO in ethanol - water mixtures and related solvent parameters. ΔG^o_{max} in kJ/mol

% EtOH	X_{EtOH}	$-\Delta G^o_{max}$ MeDPO	$-\Delta G^o_{max}$ EtDPO	$-\Delta G^o_{max}$ TPO	$E_T(30)$ kJ/mol	β_{KT}	Z	α
0	0	36.6	38.4	40.4	264.02	0.47	92.5	1.17
10	0.0329	33.7	35.4	36.2	256.49	0.465	92.0	1.14
20	0.0704	32.7	33.3	35.4	252.31	0.463	91.65	1.11
30	0.1137	30.3	30.6	32.9	246.99	0.475	91.1	1.075
40	0.1653	27.7	28.3	32.2	242.68	0.57	90.14	1.03
50	0.2277	24.8	26.8	29.9	236.83	0.64	88.9	0.98
60	0.3053	22.4	24.3	26.9	232.22	0.66	87.5	0.96
70	0.4035	20.8	23.2	25.2	228.04	0.65	86.2	0.96
80	0.5342	18.3	21.2	22.6	225.53	0.65	84.8	0.96
90	0.714	16.6	18.3	21.0	222.18	0.66	82.6	0.94
100	1	15.2	16.5	19.7	216.32	0.71	79.5	0.90

Table 4. Standard Gibbs energy of maximum adsorption of MeDPO, EtDPO and TPO in methanol - water mixtures and related solvent parameters. ΔG^o_{max} in kJ/mol

% MeOH	X_{MeOH}	$-\Delta G^o_{max}$ MeDPO	$-\Delta G^o_{max}$ EtDPO	$-\Delta G^o_{max}$ TPO	$E_T(30)$ kJ/mol	β_{KT}	Z	α
0	0.000	36.6	38.4	40.4	264.02	0.47	92.5	1.17
10	0.046	34.2	37.2	38.1	258.79	0.466	92.6	1.142
20	0.099	32.5	36.2	37.6	255.03	0.465	92.6	1.107
30	0.161	30.6	34.1	35.4	251.18	0.52	91.4	1.067
40	0.210	27.8	31.1	33.5	248.12	0.562	90.9	1.037
50	0.297	25.1	28.2	31.0	245.19	0.64	89.6	1.011
60	0.398	22.7	25.6	27.7	242.27	0.70	88.6	0.98
70	0.513	20.6	23.3	25.7	239.34	0.727	87.4	0.981
80	0.644	18.7	21.2	22.9	236.83	0.662	86.4	0.995
90	0.789	17.0	19.4	21.3	234.73	0.622	85.4	1.009
100	1	15.6	17.8	20.5	233.06	0.70	83.6	1.00

From the two terms of the right hand side of expression (10) the term $G^o(S_2)_{ads}$ was defined to represent the organic solvent- mercury interactions. At potentials corresponding to low electrode charge and particularly at the vicinity of the potential of zero charge these interactions may be assumed to be related to the ability of the organic solvent to act as electron pair donor to the mercury electrode. This assumption requires that no base electrolyte ions are present at the electrode surface, which is true in the case of $LiClO_4$ at potentials negative to -0.4V,Ag/AgCl. Therefore, the linear correlation of ΔG^o_{max} with the Kamlet - Taft parameter β_{KT}, which expresses the ability of each mixture to act as electron pair donor to the Hg electrode, may indicate the effect of this type of interactions on the overall interfacial behaviour of the system adsorbate-solvent.

In view of the monotonic variation of ΔG^o_{max} with organic solvent content, linear correlations may be mainly expected with those solvent parameters which also show a monotonic variation within a reasonable range of solvent mixture composition. In this respect the variation profile of the Kamlet - Taft parameter β_{KT} in all binary mixtures does not allow for satisfactory and physically meaningful, in terms of the corresponding interactions, linear correlations with electrochemical, kinetic and thermodynamic quantities.

The bulk solvent - solvent interactions, according to equation (11) may be summarized to the processes of hydrogen bond exchange. We assume that these processes are specifically described by a parameter expressing Lewis acidity such that the α parameter of Kamlet - Taft. For the same purpose are often also used more inclusive parameters like $E_T(30)$ and Z, which express a certain mixture of hydrogen bond ability and polarity, while Z also contains a contribution of the electron pair donicity of the solvent mixture. Good linear correlations are achieved in the case of Kamlet - Taft parameter α and also for $E_T(30)$ and Z. Such linear correlations are of the general form:

Table 5. Linear least squares correlation coefficient, R, of ΔG^{o}_{max} with parameters α, E_T and Z, in terms of eq.(12)

Solvent Mixture	MeDPO α	MeDPO $E_T(30)$	MeDPO Z	EtDPO α	EtDPO $E_T(30)$	EtDPO Z	TPO α	TPO $E_T(30)$	TPO Z
AC-H_2O	0.996 ≤90%	0.988 ≤90%	0.976 ≤80%	0.995 ≤90%	0.989 ≤90%	0.972 ≤80%	0.991 ≤90%	0.991 ≤90%	0.964 ≤80%
AN-H_2O	0.979 ≤90%	R<0.95	0.995 ≤80%	0.976 ≤90%	R<0.95	0.995 ≤80%	0.979 ≤90%	R<0.95	0.995 ≤80%
EtOH-H_2O	0.975 ≤100%	0.996 ≤100%	0.972 ≤90%	0.976 ≤100%	0.997 ≤100%	0.960 ≤90%	0.960 ≤100%	0.993 ≤100%	0.972 ≤90%
MeOH-H_2O	0.983 ≤70%	0.993 ≤100%	0.978 ≤90%	0.966 ≤70%	0.983 ≤100%	0.990 ≤90%	0.965 ≤70%	0.982 ≤100%	0.991 ≤90%

Percentage values denote the corresponding upper limit of the organic solvent volume content which was included in each correlation.

$$\Delta G^{o}_{max} = \text{const} + aP \qquad (12)$$

where P is the solvent parameter expressing the solvent effect on ΔG^{o}_{max}. The results of the above linear correlations are shown in table 5.

It can be observed that the linear dependence of ΔG^{o}_{max} on the parameters α, E_T and Z, expressing the hydrogen bond exchange ability of the solvent mixtures, is generally satisfactory, provided that in most cases the mixtures with the highest organic solvent content are not included in the linear regression. As can be easily seen from tables 1-4, the quality of these correlations does not seem to be affected by ΔG^{o}_{max} values, which are accurate and highly reproducible. It is clear that the variation profile of the solvent parameter values is the determining factor and some parameters are definitely more suitable than others for linear correlations. On the other hand it is not the specificity of the solvent parameters to particular interactions which may give better results. In this respect, it was proved that in all systems studied the use of β_{KT} parameter as a substitute of donor number to specifically express electron pair donicity of the mixtures was not successful. It may be anticipated that the more inclusive E_T and Z parameters should give better results, but this is not always true as seen from table 5. It is interesting to point out that even in the case of AN-H_2O mixtures which undergo structural changes, ΔG^{o}_{max} is well linearly correlated to Kamlet - Taft α and Kosower's Z parameters.

In conclusion, the results of the simple linear correlations of ΔG^{o}_{max} with single solvent parameters indicate that the major contribution of the mixture composition on the values of ΔG^{o}_{max} comes from the bulk solvent - solvent interactions while the impact of the interactions between the organic solvent and the Hg electrode was not found possible to be established.

The impact of each type of interactions on the overall solvent effect can be also studied by means of multilinear correlations of solvent parameter couples. When we intend to examine the effect of the solvent on a multiparameter basis then we may seek for multilinear correlations of the form:

$$\Delta G^{o}_{max} = const + aP_1 + bP_2 \tag{13}$$

where P_1 and P_2 are independent and complementary parameters [40] describing the solvent effect on ΔG^{o}_{max}. In order to examine the simultaneous effect of electron pair donicity and HB exchange ability of the organic solvent on the values of ΔG^{o}_{max}, multilinear relations of the form of equation (13) are employed. These equations correlate ΔG^{o}_{max} to the parameter couples β_{KT} - α, β_{KT} - E_T and β_{KT} - Z. According to table 6, the application of multilinear regression on equation (13) shows that good results, for all the above parameter couples, are obtained only for acetone and acetonitrile water mixtures. On the other hand satisfactory results are obtained for methanol-water mixtures for the couples β_{KT} - E_T and β_{KT} - Z and only for the couple β_{KT} - E_T in the case of ethanol-water ones.

However, it is found that a physically meaningful description of the solvent effects on the adsorption of the three phosphine oxides can be reached if we separately determine the contribution of each one of the parameters used in the multilinear regression. In this respect and in accordance to equation (13) each member of the parameter couples, namely P_1 and P_2, has a different contribution on the change of the standard Gibbs energy of adsorption. On a percentage basis these contributions can be respectively determined by means of the relations:

$$\bar{a} = \frac{100a'}{a' + b'} \tag{14}$$

$$\bar{b} = \frac{100b'}{a' + b'} \tag{15}$$

where a′ and b′ are the partial regression coefficients expressed by the following relations:

$$a' = |a| \left[\frac{\sum (P_{1,i} - \bar{P}_1)^2}{\sum (\Delta G^{o}_{max,i} - \Delta \bar{G}^{o}_{max})^2} \right]^{1/2} \tag{16}$$

$$b' = |b| \left[\frac{\sum (P_{2,i} - \bar{P}_2)^2}{\sum (\Delta G^{o}_{max,i} - \Delta \bar{G}^{o}_{max})^2} \right]^{1/2} \tag{17}$$

Table 6. Multilinear regression coefficient R^2, of ΔG^o_{max} with parameters P_1 and P_2 of solvent mixtures according to equation (13) and percentage contributions of the processes related to these parameters according to equations (14)-(17)

Solvent Mixture	Adsorbate	Equation	R^2	$\bar{a}$	$\bar{b}$
AC-H_2O	MeDPO	ΔG^o_{max}= 4.3 +13.1β_{KT} – 40.4α	0.997	11.4	85.6
≤90%		ΔG^o_{max}= 127.6 – 31.1β_{KT} – 0.57E_T	0.993	18.3	81.7
		ΔG^o_{max}= 58.0 – 36.6β_{KT} – 1.21Z	0.998	34.3	65.7
	EtDPO	ΔG^o_{max} = 3.1 +14.6β_{KT} – 41.1α	0.996	14.8	85.2
		ΔG^o_{max}= 129.0 – 30.7β_{KT} – 0.58E_T	0.993	17.7	82.3
		ΔG^o_{max}= 57.9 +38.4β_{KT} – 1.23Z	0.997	30.7	69.3
	TPO	ΔG^o_{max} = -1.7 +20.3β_{KT} – 40.7α	0.994	16.5	83.5
		ΔG^o_{max}= 122.9 –24.5β_{KT} – 0.57E_T	0.992	15.0	85.0
		ΔG^o_{max}= 52.8 +43.9β_{KT} – 1.23Z	0996	33.7	66.3
AN- H_2O	MeDPO	ΔG^o_{max}= 90.3 –52.9β_{KT} – 86.9α	0.996	20.4	79.6
≤80%		ΔG^o_{max}= 254.4 – 80.8β_{KT} – 0.96E_T	0.981	26.2	73.8
		ΔG^o_{max}= 319.9 – 20.9β_{KT} – 3.7Z	0.998	10.4	89.6
	EtDPO	ΔG^o_{max}= 90.2 –54.2β_{KT} – 87.9α	0.994	20.6	79.4
		ΔG^o_{max}= 255.3 – 82.0β_{KT} – 0.97E_T	0.976	26.3	73.7
		ΔG^o_{max}= 323.7 – 22.3β_{KT} – 3.8Z	0.999	10.2	89.8
	TPO	ΔG^o_{max}= 73.3 –46.3β_{KT} – 77.9α	0.990	20.0	80.0
		ΔG^o_{max}= 203.0 – 61.9β_{KT} – 0.81E_T	0.972	24.4	75.6
		ΔG^o_{max}= 276.6 – 15.7β_{KT} – 3.3Z	0.995	8.5	91.5
EtOH-H_2O	MeDPO	ΔG^o_{max}= 76.2 +6.3β_{KT} – 0.44E_T	0.992	8.3	91.7
≤100%	EtDPO	ΔG^o_{max}= 93.8 – 5.6β_{KT} – 0.5E_T	0.994	6.6	93.4
	TPO	ΔG^o_{max}= 100.9 – 12.6β_{KT} – 0.5E_T	0.990	13.8	86.3
MeOH-H_2O	MeDPO	ΔG^o_{max}= 129.3 + 6.7β_{KT} – 0.65E_T	0.989	9.6	90.4
≤100%		ΔG^o_{max}= 118.8 +18.6β_{KT} – 1.7Z	0.964	26.0	74.0
	EtDPO	ΔG^o_{max}= 117.3 + 10.9β_{KT} – 0.6E_T	0.970	15.7	84.3
		ΔG^o_{max}= 130.6 +16.3β_{KT} – 1.9Z	0.981	21.7	78.3
	TPO	ΔG^o_{max}= 125.1 +5.0β_{KT} – 0.64E_T	0.966	7.4	92.6
		ΔG^o_{max}= 134.1 +11.6β_{KT} – 1.9Z	0.972	16.4	83.6

The results of the application of equations (14)-(17) are provided in table 6. Solvents, like acetone and acetonitrile, which act as structure breakers of water and present positive values of entropy of mixing with water [41,42] or in other words increase the randomness of the mixture, give satisfactory results for al the types of multilinear correlations examined here.

On the contrary, solvents like methanol and ethanol, which at not very high contents enhance the structure of water [42] and are stronger donors than water, give satisfactory results in multilinear correlations involving $E_T(30)$ and Z parameters, which are sensitive with both the hydrogen bond exchange ability and the polarity of the mixtures.

The results contained in table 6 clearly show that, in what concerns the solvent effects on the phosphine oxides adsorption equilibrium, the hydrogen bond exchange bulk processes are more or less prevailing over the electron pair donation interfacial processes. Moreover, it should be noted that the plus and minus signs in the equations of table 6 respectively indicate direct and reverse proportionality of the corresponding solvent parameter and ΔG^o_{max}. In this respect, table 6 shows that in all systems studied ΔG^o_{max} of phosphine oxides is always strongly inversely proportional to the parameters α, $E_T(30)$ and Z. This means that the recession of the solvent-solvent interactions favors the establishment of adsorption equilibrium of adsorbates.

On the other hand, there is no uniform type of dependence between ΔG^o_{max} and the parameter β_{KT}, possibly because of the limited contribution of solvent – electrode interactions, which is at least statistically found. In any way, an inverse proportionality relation between ΔG^o_{max} and β_{KT}, which should be reasonably expected, is observed in the case of the strongly adsorbed aprotic solvents acetonitrile and acetone.

CONCLUSION

As a consequence of the assumptions made in this chapter, i.e. (a) the heterosolvation, (b) the weak interactions of the adsorbates with both water and the organic solvents used and (c) the independence of adsorbate - electrode interactions from the mixtures' composition, the solvent effects on ΔG^o_{max} are related only to the organic solvent-electrode interactions and to all kinds of solvent - solvent interactions affecting both the overall and the local structure of the mixture.

The application of multilinear correlations in combination with the determination of the relative importance of solvent acidity and basicity, by means of relations (13) -(17), brought out the prevalence of solvent - solvent interactions and the limited role of solvent - electrode interactions. This difference is particularly clear for solvents acting as structure makers in the aqueous mixture like methanol and ethanol. In these mixtures the effect of solvent - solvent interactions is better expressed by parameters combining hydrogen bond exchange ability and polarity, that is the $E_T(30)$ and Z parameters. In aqueous mixtures of aprotic solvents, acetone and acetonitrile, acting as structure breakers of water the multilinear relations also include the Kamlet - Taft parameter α, which is non sensitive to polarity, while the prevalence of solvent-solvent interactions is usually not particularly strong.

ACKNOWLEDGEMENT

The author wishes to express his thanks to Professor Robert de Levie for donating his electrochemical impedance equipment to the laboratory of Physical Chemistry of the University of Thessaloniki.

REFERENCES

[1] Conway, B.E. *J. Solution Chem.*, 1978, *7*,721-770.
[2] Conway, B.E. *Electrochim. Acta*, 1995, *40*,1501-1512.
[3] Conway, B.E. *Solid State Ionics*, 1997, *94*,165-170.
[4] Conway, B.E. ; Ayranci, E. *J. Solution Chem.*, 1999, *28*,163-192.
[5] Fawcett, W.R.; Motheo, A.J. *Electrochim. Acta*, 1991,*36*,1971-1977.
[6] Levi, M.D.; Slepakov, A.V.; Damaskin, B.B.; Bagotskaya, I.A. *J.Electroanal. Chem.*, 1982,*138*,1-27.
[7] Trasatti, S. *J.Electroanal.Chem.*, 1975,*65*,815-829.
[8] Reichardt, C. *Solvents and Solvent Effects in Organic Chemistry*, VCH, New York, 1990,pp 339-392.
[9] Marcus, Y. *The Properties of Solvents*, John Wiley & Sons, New York, 1999, pp131-199.
[10] Damaskin, B.B. *Russian J. of Electrochem.*, 1994,*30*,91-94.
[11] Anastopoulos, A.G. *Z.Phys.Chem.*, 2002,*216*,1095-1106.
[12] Anastopoulos, A.G. *Electrochim.Acta*, 1996,*41*,2537-2544.
[13] Anastopoulos, A.; Christodoulou, A.; Jannakoudakis, D. *Z.Phys.Chem.*, NF, 1983,*137*,231-246.
[14] Anastopoulos, A.; Christodoulou,A. *Coll. Czech. Chem. Commun.*, 1988, *53*, 732-742.
[15] Anastopoulos, A.; Christodoulou,A. ; Moumtzis, I. *Can.J.Chem.*, 66(1988) 1053-1058.
[16] Anastopoulos, A.; Moumtzis, I. *J.Electroanal.Chem.*, 1990, *294*,143-150.
[17] Anastopoulos, A.G. *J.Phys.Chem.*B, 2000,*104*, 5102-5109.
[18] Anastopoulos, A.G.; Hadjiloizou, P. ; Nikoletou, D. *Langmuir*, 2000,*16*, 7789-7795.
[19] Anastopoulos, A.G. ; Nikolaidis, G.N. *J.Electroanal.Chem.*, 2004, *571*,309-317.
[20] Marcus, Y. *J.Chem.Soc.Perkin Trans.*II, 1994,1751-1758.
[21] Kalidas, C.; Hefter, G.; Marcus, Y. *Chem.Rev.*, 2000,*100*,819-852.
[22] Cheong, W.J.; Carr, P.W. *Anal.Chem.*, 1988,*60*,820-826.
[23] Wrona, P.K.; Krygowski, T.M.; Zielkowska, U. *Z.Naturforsch.*, 1989, 44b, 673-678
[24] Maksymiuk, K.; Stroka, J.; Galus, Z. *J.Electroanal.Chem.*, 1984,*181*,51-63.
[25] Gutmann, V. *Coord.Chem.Rev.*, 1976,*18*,225-255.
[26] G.Gritzner, G. *J.Phys.Chem.*, 1986,*90*,5478-5485.
[27] Reichardt, C. *Chem.Rev.*, 1994,*94*, 2319-2358.
[28] Izutsu, K. *Electrochemistry in Nonaqueous Solutions*, Wiley-VCH, Weinheim, 2002,pp13-50.
[29] Kosower,E.M. *J.Am.Chem.Soc.*, 1958,*80*,3253-3260.
[30] Grahame, D.C. *J.Am.Chem.Soc.*, 1949,*71*,2975-2978.

[31] Nikitas, P.; Pappa-Louisi, A.; Jannakoudakis, D. *J.Electroanal.Chem.*, 1984,*162*,175-187.

[32] De Battisti, A.; Trasatti, S. *J.Electroanal.Chem.*, 1973,*48*,213-238,

[33] Borkowska, Z. *J.Electroanal.Chem.*, 1975,*63*, 379-392.

[34] Taraszewska, J. *J.Electroanal.Chem.*, 1974,*49*,443-451.

[35] Anastopoulos, A.G. ; Latsinoglou, Th. *J.Solution Chem.*, 2005,*34*,1341-1356.

[36] Petrella, G.; Petrella, M. *Electrochim Acta*, 1982,*27*,1733-1739.

[37] Bryant, P.J.R.; Wardrop, A.W.H. *J.Chem.Soc.*, 1957,895-906.

[38] Marcus, Y.; Migron, Y. *J.Phys.Chem.*, 1991,*95*,400-406.

[39] Wrobel, R.; Makowski, M.; Czaja, M.; Chmurzynski, L. *Polish. J. Chem.*, 2001,*75*,1767-1774.

[40] Krygowski, T.M.; Fawcett, W.R. *J.Am.Chem.Soc.*, 1975,*97*,2143-2148.

[41] Venables, D.S.; Schmuttenmaer, C.A. *J.Chem.Phys.*, 1998,*108*,4935-4944.

[42] Suresh, S.J.; Naik, V.M. *J.Chem.Phys.*, 2002,*116*,4212-4220.

In: Solution Chemistry Research Progress
Editor: Darian V. Bostrelli, pp. 133-148
ISBN 978-1-60021-993-1

Chapter 6

The Boiling Point Elevations of Electrolyte Solutions. Activity and Osmotic Coefficients at the Boiling Point Temperatures.

Alexander Apelblat*

Chemical Engineering Department, Ben Gurion University of the Negev,
Beer Sheva, Israel

Abstract

Using data available in the literature (mainly from the Timmermans tabulation), the experimental boiling points of electrolyte solutions at constant pressures are expressed as a function of the concentration of the dissolved electrolyte in water. The elevations of boiling points are used in the evaluation of the Pitzer parameters of the activity and osmotic coefficients at the boiling point temperatures for 67 electrolytes of the 1:1, 1:2, 2:1 and 2:2 types.

Introduction

Osmotic and activity coefficients of electrolytes can be determined from colligative properties of solutions. The colligative property depends only on the activity of the solvent and not on the nature of the solute present in the solution. If the activity of the solvent is a measurable quantity, then the activity coefficient of the solute can be derived from the integration of the Gibbs-Duhem equation. Ebullioscopy is the colligative property associated with the elevation of the boiling temperature of the solvent when the solute is non-volatile. Usually when measuring activities, other colligative properties of electrolyte solutions, such as the freezing point depression (cryoscopy), the lowering of the solvent's vapour pressure and the osmotic pressure, are preferred. In principle, the determination of the boiling point

* apelblat@bgu.ac.il

elevation is similar to that of the freezing point depression. However this method is less frequently used because the ebullioscopic constant is about four to five times smaller than the corresponding cryoscopic constant and the boiling point depends strongly on the imposed pressure. Thus, considering desired precision and complexity of measurements, it is generally accepted that the ebullioscopic method is rather suitable at higher temperatures, above or near the normal boiling point. However, since the solubilities of electrolytes usually increase with temperature increasing, the investigated concentration range is greater than in other methods. It is worthwhile to note that the measurement of the boiling temperature elevations at a constant pressure is equivalent to the determination of the vapour changes at constant temperature. Ebullioscopic experiments made at different pressures allow for the determination of activity data over large temperature intervals.

The major contributions to our knowledge regarding the boiling temperatures of electrolyte solutions are from measurements performed by Gerlach (1887) [1], Baroni {1893) [2], Schlamp (1894) [3], Buchanan (1899) [4], Kahlenberg (1901) [5], Johnston (1906) [6] and Jablczynski and Kon (1923) [7]. The ebullioscopic method was also popular during the 1930's and 40's with the outstanding contribution of Swietoslawski and his school [8] to the experimental aspects of the method and that of Smith and coworkers [9-12] concerning the thermal properties of water at boiling points at various pressures. A number of summaries can be found in the literature dealing with detailed descriptions of ebulliometers, their construction and applied procedures [13-16]. In modern determinations of the solvent activities, there is increased tendency to replace the ebullioscopic method by the high-temperature isopiestic technique.

The knowledge of solubilities at high temperatures in itself is of considerable importance in chemical engineering practice, e.g. in the design and control of industrial evaporators. Therefore the majority of earlier investigations were associated with the solubilities rather than with the colligative properties of solutions. Many of these earlier solubility investigations can be considered, by present-day standards, as not especially accurate as the boiling of concentrated solutions is frequently accompanied with the superheating of the solutions. On the other hand, precise ebullioscopic measurements were usually linked with the determination of the van't Hoff parameters and finding molecular masses of substances.

There are several tabulations dealing with different properties of aqueous electrolyte solutions. In the particular case of boiling temperatures, the most complete compilation of experimental data is that of Timmermans [17,18] which uses original literature publications and is easier to use than the International Critical Tables [19] or the Landolt-Börnstein tabulations [20]. In Timmermans compilations there is a uniform presentation of the boiling curve – the boiling temperatures versus concentration which is usually expressed in the weight percent units. All the available sets of experimental points are presented separately, but the exact imposed pressure is not always given implying that the imposed pressure is the atmospheric pressure. The Timmermans compilations cover the period before 1959, but the ebullioscopic investigations after that year are rather rare. It is also possible to find in various handbooks the temperature dependence of electrolyte solubilities. But these solubilities are often interpolated and smoothed. Prediction and estimation procedures and various forms of correlation of boiling temperatures are discussed in a detail by Horvath [21] (for the estimation of osmotic and activity coefficient at elevated temperatures see Apelblat [22]). The elevation of boiling points in saturated solutions of electrolytes was correlated by Meranda and Furter [23].

Thermodynamic Relations Associated with the Boiling–Point Elevation of Aqueous Electrolyte Solutions

For a vapour – liquid equilibrium system in which the solvent is the only volatile component, let us denote the solvent by the subscript 1 and the solute by the subscript 2. At the boiling point of a solution of molality m, $T_{b.p.}(m) \equiv T$, the chemical potential of the solvent in the liquid phase is equal to that of the pure solvent. The chemical potential of the solvent in the solution can be expressed in terms of its activity as

$$\mu_1(T,P,m) = \mu_1^0(T,P) + RT\ln a_1 \tag{1}$$

If the reference state for both phases is the same, i.e. the pure liquid solvent, the change in the solvent activity with temperature T at constant P and m is [13]

$$d\ln a_1 = \left(\frac{\partial \ln a_1}{\partial T}\right)_{P,m} dT \tag{2}$$

where

$$\left(\frac{\partial \ln a_1}{\partial T}\right)_{P,m} = -\frac{(\bar{H}_1 - \bar{H}_1^0)}{RT^2} = -\frac{\bar{L}_1}{RT^2} \tag{3}$$

$\bar{H}_1 - \bar{H}_1^0 \equiv \bar{L}_1$ denotes the relative partial molar heat content of the solvent. Differentiation of $\bar{L}_1$ with respect to temperature can be expressed in terms of the partial molar heat capacity $\bar{J}_1$

$$\left(\frac{\partial \bar{L}_1}{\partial T}\right)_{P,m} = \bar{J}_1 = \bar{C}_{P1} - \bar{C}_{P1}^0 \tag{4}$$

If the pressure P is the vapour pressure of the pure solvent (water) then $P = P_w$ at temperature $T_{b.p.}(m = 0) = T_w$ (the boiling point of pure water) and $\bar{L}_1$ in Eq.(3) is the molar enthalpy of vaporization. The elevation of the boiling point of the aqueous solution having molality m and under pressure P is denoted by $\theta(m)$

$$\theta(m) = \Delta T = T_{b.p.}(m) - T_w \tag{5}$$

If $\bar{L}_1$ and $\bar{J}_1$ are expressed as polynomials of θ, based on the thermal properties of water determined by Johnson and Smith [12], the integration of Eq.(3) gives the activity of water [13]

$$\ln a_1(m;T) = -\frac{M_1}{1000\lambda(T)}\sum_{n=1}^{4} C_n(T)\theta(m)^n \tag{6}$$

where M_1 is the molar mass of water, $T \equiv T_{b.p.}(m)$ and the numerical coefficients $C_n(T)$ are:

$$C_1(T) = 1 \tag{7a}$$

$$C_2(T)\cdot 10^3 = -3.1602 + 6.988\cdot 10^{-3}(T/K - 373.15) - 2.518\cdot 10^{-3}(T/K - 373.15)^2 \tag{7b}$$

$$C_3(T)\cdot 10^6 = 7.8424 - 5.001\cdot 10^{-2}(T/K - 373.15) + 5.765\cdot 10^{-4}(T/K - 373.15)^2 \tag{7c}$$

$$C_4(T)\cdot 10^8 = -1.8440 + 2.047\cdot 10^{-2}(T/K - 373.15) - 3.850\cdot 10^{-4}(T/K - 373.15)^2 \tag{7d}$$

The ebullioscopic constants, $\lambda\,(T)$, for water are:

$$\lambda(T)/\mathrm{K}\cdot\mathrm{mol}^{-1} = 0.51276 + 3.335\cdot 10^{-3}(T/K - 373.15) + 7.326\cdot 10^{-6}(T/K - 373.15)^2 \tag{8}$$

Thus, the ebullioscopic constant of water at its normal boiling point (T = 373.15 K and P = 760 mmHg = 101.325 kPa) is 0.51276 K mol^{-1}. The boiling point of water under pressure P is given by:

$$T/\mathrm{K} = 373.15 + 0.27645(P/\mathrm{kPa}\text{-}101.325)\text{-}1.13413\cdot 10^{-3}(P/\mathrm{kPa}\text{-}101.325)^2 + 6.84303\cdot 10^{-6}(P/\mathrm{kPa}\text{-}101.325)^3 \tag{9}$$

The osmotic coefficient of electrolyte is defined by:

$$\phi(m,T) = -\frac{1000}{\nu M_1 m}\ln a_1(m,T) \tag{10}$$

where $\nu = \nu_+ + \nu_-$ are the numbers of cations and anions produced by the dissociation of one molecule of electrolyte and M_1 is the molar mass of the solvent. Introducing Eq.(6) into Eq.(9) we have:

$$\phi(m;T) = \frac{1}{\nu m\lambda(T)}\sum_{n=1}^{4} C_n(T)\theta(m)^n \tag{11}$$

The osmotic coefficients calculated from Eq.(11) can be fitted to the Pitzer equation with three adjustable constants $\beta^{(0)}$, $\beta^{(1)}$ and C^{ϕ} (valid for all electrolytes with the exception of the 2:2 type) [24]

$$\phi = 1 + |z_+ z_-| f^{\phi} + 2\frac{\nu_+ \nu_-}{\nu} B^{\phi} m + 2\frac{(\nu_+ \nu_-)^{3/2}}{\nu} C^{\phi} m^2 \tag{12a}$$

$$f^{\phi} = -\frac{A_{\phi}(T)\sqrt{I}}{1 + 1.2\sqrt{I}} \tag{12b}$$

$$B^{\phi} = [\beta^{(0)} + \beta^{(1)} e^{-2\sqrt{I}}] \tag{12c}$$

The corresponding activity coefficients are:

$$\ln \gamma_{\pm} = |z_+ z_-| f^{\gamma} + 2\frac{\nu_+ \nu_-}{\nu} B^{\gamma} m + 2\frac{(\nu_+ \nu_-)^{3/2}}{\nu} C^{\gamma} m^2 \tag{13a}$$

$$f^{\gamma} = -A_{\phi}(T)[\frac{\sqrt{I}}{1 + 1.2\sqrt{I}} + \frac{2}{1.2}\ln(1 + 1.2\sqrt{I})] \tag{13b}$$

$$B^{\gamma} = 2\beta^{(0)} + \frac{\beta^{(1)}}{2I}[1 - (1 + 2\sqrt{I} - 2I)e^{-2\sqrt{I}}] \tag{13c}$$

$$C^{\gamma} = \frac{3}{2} C^{\phi} \tag{13d}$$

where I denotes the ionic strength which in the case of a single electrolyte takes the form

$$I = \frac{1}{2}(z_+^2 \nu_+ + z_-^2 \nu_-) m \tag{14}$$

and $A_{\phi}(T)$ is the Debye-Hückel constant

$$A_{\phi}(T) = 0.4603 + 1.187 \cdot 10^{-3}(T/K - 373.15) + 3.568 \cdot 10^{-6}(T/K - 373.15)^2 \tag{15}$$

In the particular case of a 2:2 electrolytes, the additional parameter, $\beta^{(2)}$, is introduced to the Pitzer equations and [24]

$$\phi = 1 + 4 f^{\phi} + B^{\phi} m + C^{\phi} m^2 \tag{16a}$$

$$f^{\phi} = -\frac{A_{\phi}(T)\sqrt{I}}{1+1.2\sqrt{I}} \tag{16b}$$

$$B^{\phi} = [\beta^{(0)} + \beta^{(1)} e^{-1.4\sqrt{I}} + \beta^{(2)} e^{-12\sqrt{I}}] \tag{16c}$$

and

$$\ln \gamma_{\pm} = 4 f^{\gamma} + B^{\gamma} m + C^{\gamma} m^2 \tag{17a}$$

$$f^{\gamma} = -A_{\phi}(T)[\frac{\sqrt{I}}{1+1.2\sqrt{I}} + \frac{2}{1.2}\ln(1+1.2\sqrt{I})] \tag{17b}$$

$$\begin{aligned} B^{\gamma} &= 2\beta^{(0)} + \frac{2\beta^{(1)}}{1.4^2 I}[1-(1+1.4\sqrt{I} - \frac{1.4^2}{2} I)e^{-1.4\sqrt{I}}] \\ &+ \frac{2\beta^{(2)}}{12^2 I}[1-(1+12\sqrt{I} - \frac{12^2}{2} I)e^{-12\sqrt{I}}] \end{aligned} \tag{17c}$$

$$C^{\gamma} = \frac{3}{2} C^{\phi} \tag{17d}$$

There is an alternative method to describe the colligative properties of electrolyte solutions. At the boiling point temperature T, the vapour pressure of water above the solution is equal to that of the external pressure P. This vapour pressure is lower than that of pure water at T and the difference in these vapour pressures

$$\Delta P(m;T) = P_1^0(T) - P(m;T) = [1 - a_1(m;T)]P_1^0(T) \tag{18}$$

is called the vapour pressure lowering of electrolyte solution. The values of vapour pressures of the pure water are available from the Saul and Wagner equation [25]

$$\begin{aligned} \ln\left(\frac{P_1^0(T)}{P_{crit.}}\right) &= \frac{\tau}{1-\tau}[-7.85823 + 1.8399\tau^{1/2} - 11.7811\tau^2 + 22.6705\tau^{5/2} - \\ &15.9393\tau^3 + 1.77516\tau^{6.5}] \end{aligned} \tag{19}$$

$$\tau = 1 - \frac{T}{T_{crit.}} \quad ; \quad T_{crit.} = 647.14\text{K}; \quad P_{crit.} = 2.2064 \cdot 10^4 \text{kPa}.$$

Using this equation, it is possible to correlate ΔP with the boiling point elevation $\theta(m)$. Under atmospheric pressure, $P = 101.325$ kPa, $\Delta P = f(\theta)$ can be approximated by

$$\Delta P/\mathrm{kPa} = 3.6298(\theta/\mathrm{K}) + 0.05089(\theta/\mathrm{K})^2 + 0.0005209(\theta/\mathrm{K})^3 \quad (20)$$

In terms of the vapour pressure lowering, the osmotic coefficient of the electrolyte is

$$\phi(m;T) = \frac{55.508}{\nu m}\left[\ln\left(1 - \frac{\Delta P(m,T)}{P_1^0(T)}\right) - \frac{\mathrm{B}_1(T)\Delta P(m,T)}{RT}\right] \quad (21)$$

where R = 8314.7 $cm^3 \cdot kPa \cdot K^{-1} \cdot mol^{-1}$ is the gas constant and $B_1(T)$ is the second virial coefficient of water vapours [26]

$$\mathrm{B}_1(T)/\mathrm{cm}^3 \cdot \mathrm{mol}^{-1} = -11766 + 66.962(T/\mathrm{K}) - 0.12993(T/\mathrm{K})^2 + 0.00008477(T/\mathrm{K})^3 \quad (22)$$

As has been shown by the author [22], at constant molality m, the vapour pressure lowering of electrolyte solutions, $\Delta P(m,T)$, over a wide temperature range, can be very accurately computed by the following equation

$$\ln[\Delta P(m,T)] = \mathrm{A} + \frac{\mathrm{B}}{(T/\mathrm{K})} + \mathrm{C}[\ln(T/\mathrm{K}) - \frac{T}{2T_{crit.}}] \quad (23)$$

where A, B and C are adjustable constants and $T_{crit.}$ is the critical temperature of pure water. Thus, in principle, the knowledge of these constants at three different molalities allows the evaluation of the coefficients in the Pitzer equations and also to estimate the osmotic and activity coefficients of an electrolyte at any T and m.

DATA ANALYSIS, OPTIMIZATION PROCEDURES AND RESULTS

The experimental values of the boiling point elevations $\theta(m)$ (compiled in the Timmermans tabulation) are mostly reported from one investigation only. Some systems (NaCl or KCl for example) were studied repeatedly, but often results either disagree or cover different concentration ranges. In some cases, the ebullioscopic measurements were performed for the same system but at different external pressures. The accuracy of the reported results was never estimated in the original papers, and it probably varies significantly from one paper to another, depending strongly on the equipment used and the investigated concentration range. This varied accuracy poses significant numerical difficulties in the optimization procedure to calculate the osmotic coefficients in the Pitzer equations. Whereas in dilute solutions, the modified Debye-Hückel term of the Pitzer equations correctly represents the experimental data, inclusion of the dilute data in the optimization at higher concentrations leads to mathematical problems. An example of this was observed when the investigation of Plake [27], where $\theta(m)$ values in very dilute solutions (0.0008 $mol\cdot kg^{-1}$ to 0.3 $mol\cdot kg^{-1}$) for many 1:2, 2:1 and 2:2 electrolytes, were combined with those of more concentrated solutions. Hence, arbitrarily for higher concentration, the values of $\theta(m)$ from only moderately concentrated solutions of electrolytes ($m > 0.5$ $mol\cdot kg^{-1}$) were preferred in the calculations. The chosen concentrations m (denoted as m_{max} in Tables 1-6) were always greater or equal than that used in the calculations of the Pitzer coefficients, but at the same time were also always considerably below that of the saturated solutions as the solubilities in water of many electrolytes at high temperatures can be very large. This selective range of m_{max} values was used to properly present the functional dependence of osmotic and activity coefficients with concentration m. The agreement with the experimental results is expressed in terms of the corresponding standard deviations $\sigma(\theta)$ and $\sigma(\phi)$. For the 2:2 electrolytes, the last exponential term in Eq.(16c) is only important in very dilute solutions [24]. Therefore in the optimization procedure, the coefficients $\beta^{(2)}$ were assumed to be the same as those used by Pitzer at 298.15 K.

Table 1. Boiling point elevations (equation (24)) for 1:1 electrolytes

	P/kPa	T/K	A	B	C	m_{max}/m^0	$\sigma(\theta)$
HF	101.325	373.150	0.8194	-1.0064	-1.0064	1.0	0.016
HCl	101.325	373.150	0.9400	0.0839	0.0839	2.1	0.001
$HClO_3$	101.325	373.150	0.7163	-1.7128	-1.7128	0.7	0.007
NH_4Cl	73.374	364.388	0.7685	0.0359	0.0359	14.0	0.135
	82.660	367.550	0.8481	0.0185	0.0185	14.8	0.060
	99.012	372.504	0.8369	0.0245	0.0245	14.7	0.098
	101.325	373.150	0.7537	0.0366	0.0366	16.3	0.051
NH_4Br	100.125	372.817	0.9194	0.0141	0.0141	2.0	0.002
NH_4I	101.058	373.076	0.9393	-0.0083	-0.0083	1.5	0.002
NH_4NO_3	101.325	373.150	0.8078	-0.0048	-0.0048	31.0	0.040
LiCl	101.325	373.150	0.7486	0.1284	0.1284	9.7	0.080
LiBr	101.325	373.150	1.1411	-0.0561	-0.0561	4.8	0.071
LiI	101.325	373.150	0.7448	0.0769	0.0769	9.9	0.032
LiOH	101.325	373.150	1.0247	-0.0148	-0.0148	10.0	0.001
$LiNO_3$	100.000	372.782	0.8545	0.1096	0.1096	8.1	0.046
$LiC_2H_3O_2$	100.000	372.782	0.8747	0.0978	0.0978	6.4	0.046
$LiC_7H_5O_3$	101.325	373.150	0.9109	0.0726	0.0726	2.4	0.011
NaCl	99.125	372.536	1.0403	0.0298	0.0298	3.4	0.031
	99.139	372.540	0.7260	0.1464	0.1464	5.4	0.046
	99.592	372.667	0.9787	0.0886	0.0886	3.5	0.003
	100.432	372.902	0.9930	0.0316	0.0316	3.7	0.027
	100.525	372.928	0.8884	0.0636	0.0636	2.0	0.004
	100.925	373.039	0.9292	0.0325	0.0325	2.3	0.002
	101.325	373.150	0.84257	0.06510	0.0651	4.4	0.009

NaBr	99.565	372.660	0.9085	0.0932	0.0932	2.2	0.002
	101.325	373.150	0.8904	0.0921	0.0921	3.9	0.029
NaI	100.752	372.991	0.9424	0.0994	0.0994	2.0	0.001
	101.325	373.150	0.8918	0.1177	0.1177	8.1	0.063
$NaNO_3$	101.325	373.150	0.9681	-0.0168	-0.0168	10.4	0.048
NaOH	101.325	373.150	0.8278	0.0838	0.0838	26.6	0.250
NaSCN	101.325	373.150	0.9434	0.0463	0.0463	4.3	0.006
$NaC_2H_3O_2$	100.000	372.782	0.8631	0.0985	0.0985	7.6	0.041
	101.325	373.150	0.8893	0.0647	0.0647	8.1	0.031
$NaC_3H_5O_3$	98.939	372.484	1.0068	0.0069	0.0069	20.8	0.135
KCl	98.152	372.261	0.8836	0.0331	0.0331	6.6	0.032
	98.339	372.314	0.9883	0.0396	0.0396	4.3	0.005
	98.725	372.424	0.9229	0.0050	0.0050	1.5	0.001
	99.192	372.555	0.9960	0.0324	0.0324	4.1	0.032
	100.445	372.906	0.8564	0.0261	0.0261	2.0	0.023
	101.325	373.150	0.8740	0.0410	0.0410	7.8	0.006
KBr	98.592	372.386	0.8570	0.0724	0.0724	4.1	0.014
	98.685	372.412	0.9233	0.0375	0.0375	4.3	0.007
	99.538	372.652	0.9201	0.0325	0.0325	1.4	0.001
	101.325	373.150	0.9093	0.0368	0.0368	5.6	0.002
KI	98.132	372.255	0.9031	0.0778	0.0778	2.9	0.004
	98.659	372.405	0.8259	0.1090	0.1090	6.3	0.029
	99.565	372.660	0.8912	0.1107	0.1107	1.3	0.001
	101.325	373.150	0.9546	0.0376	0.0376	8.7	0.083
$KClO_3$	73.374	364.388	0.6499	0.0853	0.0853	4.2	0.076
	82.767	367.585	0.8208	-0.0256	-0.0256	4.2	0.033
	96.992	371.930	0.8630	-0.0283	-0.0283	4.7	0.005
	98.445	372.344	1.0880	-0.1389	-0.1389	4.0	0.011
	98.845	372.457	0.8909	-0.0400	-0.0400	3.9	0.046
	101.325	373.150	1.5016	-0.2572	-0.2572	2.9	0.013
	102.418	373.451	1.1380	-0.1008	-0.1008	5.0	0.037
KNO_3	98.819	372.450	0.8859	-0.0398	-0.0398	6.2	0.004
	99.285	372.581	0.8837	-0.0321	-0.0321	7.0	0.026
	99.765	372.716	0.7941	0.0138	0.0138	3.3	0.026
	101.325	373.150	0.9378	-0.1344	-0.1344	1.7	0.002
KOH	101.325	373.150	1.3209	0.0113	0.0113	17.8	0.109
$KC_2H_3O_2$	101.325	373.150	1.6224	-0.0128	-0.0128	19.9	0.024
RbCl	99.605	372.671	0.8468	0.0390	0.0390	5.0	0.013
CsCl	99.532	372.651	0.8898	0.0062	0.0062	5.3	0.029
$CsNO_3$	99.179	372.551	1.1020	-0.2937	-0.2937	2.7	0.038
$AgNO_3$	98.472	372.352	0.8799	-0.0716	-0.0716	8.0	0.053
	100.858	373.021	0.8403	-0.0654	-0.0654	5.0	0.012
	101.325	373.150	0.6087	-0.0117	-0.0117	7.7	0.360

Table 2. Coefficients of the Pitzer equations (16) for 1:1 electrolytes

	A_ϕ	$\beta^{(0)}$	$\beta^{(1)}$	C^ϕ	m_{max}/m^0	$\sigma(\phi)$	Ref.
HF	0.4603	0.5687	-4.4568	-0.3552	1.0	0.028	31,32
HCl	0.4603	0.1517	0.4783	-0.0076	2.1	0.001	7
$HClO_3$	0.4603	-0.8855	-3.3547	1.1266	0.7	0.001	28
NH_4Cl	0.4502	0.0465	0.2926	-0.0023	14.0	0.001	4
	0.4538	0.0396	0.6041	-0.0018	14.8	0.003	4
	0.4595	0.0418	0.2584	-0.0019	14.7	0.004	4
	0.4603	0.0437	-0.4173	-0.0022	16.3	0.009	1
NH_4Br	0.4599	0.0914	0.3105	-0.0128	2.0	0.001	7
NH_4I	0.4602	0.0554	0.5068	0.0068	1.5	0.001	7
NH_4NO_3	0.4603	0.0108	-0.0847	-0.0007	19.7	0.013	1
LiCl	0.4603	0.2262	-1.1431	-0.0258	4.7	0.030	3,6
LiBr	0.4603	0.0094	1.9178	0.0293	4.8	0.019	6
LiI	0.4603	0.3230	-2.2849	-0.0556	4.6	0.032	6
LiOH	0.4603	0.0422	1.1229	-0.0029	10.0	0.019	41
$LiNO_3$	0.4599	0.1534	-0.0633	-0.0097	8.1	0.013	39
$LiC_2H_3O_2$	0.4599	0.1488	0.0588	-0.0095	6.4	0.008	38
$LiC_7H_5O_3$	0.4603	0.3241	-0.4173	-0.1862	2.4	0.018	3
NaCl	0.4596	0.0510	1.4757	0.0192	3.4	0.006	5
	0.4596	0.1533	-0.5635	-0.0099	5.4	0.001	5
	0.4597	0.0986	1.0706	0.0065	3.5	0.009	5
	0.4600	0.0701	1.0236	0.0327	3.7	0.004	4
	0.4600	0.3193	-0.5322	-0.1049	1.5	0.006	5
	0.4602	0.1352	0.2639	-0.0179	2.3	0.004	7
	0.4603	0.1594	-0.4379	-0.0147	4.4	0.011	34
NaBr	0.4597	0.1822	0.1695	-0.0218	2.2	0.019	7
	0.4603	0.1447	0.2284	-0.0108	2.2	0.007	31
NaI	0.4601	0.1588	0.5365	-0.0054	2.0	0.001	7
	0.4603	0.1421	0.5288	-0.0029	6.9	0.005	3,6,34
$NaNO_3$	0.4603	0.0244	0.8444	-0.0013	10.4	0.007	33
NaOH	0.4603	0.0913	0.3498	-0.0024	26.6	0.022	1
NaSCN	0.4603	0.0990	0.5710	-0.0007	4.3	0.001	34
$NaC_2H_3O_2$	0.4599	0.1600	-0.1440	-0.0139	6.9	0.018	40
	0.4603	0.1008	0.4004	-0.0058	6.6	0.007	1,3
$NaC_3H_5O_3$	0.4595	0.0531	1.1304	-0.0025	13.4	0.008	35
KCl	0.4592	0.0778	0.2266	-0.0032	6.6	0.004	5
	0.4593	0.0626	1.1158	0.0043	4.3	0.009	5
	0.4594	0.0934	0.3225	-0.0149	1.5	0.001	7
	0.4596	0.0436	1.2187	0.0090	4.1	0.012	5
	0.4600	2.3423	-5.3208	-1.8208	0.8	0.033	37

	0.4603	0.0882	0.0565	-0.0061	7.8	0.031	2
KBr	0.4594	0.1815	-0.3699	-0.0224	4.1	0.016	5
	0.4594	0.1147	0.3125	-0.0101	4.3	0.008	5
	0.4597	0.1368	0.2369	-0.0283	1.4	0.001	7
	0.4603	0.0940	0.2927	-0.0061	5.6	0.009	5
KI	0.4592	0.2356	-0.1559	-0.0452	2.9	0.016	5
	0.4594	0.2188	-0.7171	-0.0236	6.3	0.039	5
	0.4597	0.2531	-0.1166	-0.0723	1.3	0.001	7
	0.4603	0.0859	0.6745	-0.0039	8.7	0.004	1,6
$KClO_3$	0.4502	0.0901	-0.8101	-0.0183	4.2	0.000	4
	0.4538	0.0126	0.0269	-0.0016	4.2	0.001	4
	0.4589	0.0091	0.1988	-0.0021	4.7	0.001	4
	0.4593	-0.1749	2.2084	0.0432	3.5	0.025	5
	0.4595	0.0387	0.0899	-0.0085	3.9	0.005	4
	0.4603	-0.2489	4.7554	0.0799	2.9	0.008	2
	0.4607	-0.0142	1.6744	0.0015	5.0	0.000	4
KNO_3	0.4595	0.0394	-0.0235	-0.0084	6.2	0.020	5
	0.4596	0.0263	0.0942	-0.0057	7.0	0.013	5
	0.4598	0.1300	-0.7789	-0.0383	2.7	0.005	4
	0.4603	-0.0114	0.3024	0.0003	1.7	0.003	2
KOH	0.4603	0.1035	2.9782	-0.0013	17.8	0.029	2
$KC_2H_3O_2$	0.4603	0.1407	4.5350	-0.0080	10.7	0.012	1
RbCl	0.4597	0.1306	-0.3202	-0.0188	3.9	0.010	4
CsCl	0.4597	0.0741	0.1124	-0.0043	5.3	0.008	4
$CsNO_3$	0.4596	-0.3777	2.4774	0.1512	2.7	0.013	6
$AgNO_3$	0.4594	-0.0195	0.1220	-0.0008	8.0	0.013	5
	0.4601	0.0035	-0.2539	-0.0062	5.0	0.009	4
	0.4603	-0.0200	-1.4147	0.0003	7.7	0.064	36

The boiling point elevations θ (m) were calculated using the third order polynomial

$$\theta(m)/\mathrm{K} = \mathrm{A}(m/m^0) + \mathrm{B}(m/m^0)^2 + \mathrm{C}(m/m^0)^3 \qquad (24)$$
$$m^0 = 1 \cdot \mathrm{mol} \cdot \mathrm{kg}^{-1}$$

The coefficients A, B and C were evaluated by an unweighted multivariate least-square method and they are reported in Table 1 for 1:1 electrolytes, Table 3 for 1:2 and 2:1 electrolytes and Table 5 for 2:2 electrolytes. Using the boiling point elevations θ(m), the osmotic coefficients were calculated (Eqs.(6)-(11)) and expressed by the Pitzer equations (16) for the 1:1, 1:2 and 2:1 electrolytes and equations (17) for the 2:2 electrolytes. The Pitzer coefficients $\beta^{(0)}$, $\beta^{(1)}$ and C^{ϕ} were determined using the same unweighted multivariate least-square method, and they are reported in Table 2 for the 1:1 electrolytes, Table 4 for the 1:2 and 2:1 electrolytes and Table 6 for the 2:2 electrolytes.

Table 3. Boiling point elevations (equation (24)) for 1:2 and 2:1 electrolytes

	P/kPa	T/K	A	B	C	m_{max}/m^0	σ(θ)
$(NH_4)_2SO_4$	73.242	364.341	0.4017	0.1736	-0.0145	8.8	0.031
	82.704	367.565	0.6910	0.0819	-0.0070	8.1	0.066
	99.179	372.551	0.8589	0.0152	0.00018	7.3	0.030
	101.325	373.150	0.7712	0.0487	-0.0035	8.7	0.020
	102.278	373.413	0.9799	-0.0184	0.00246	7.5	0.054
Li_2SO_4	101.325	373.150	0.8015	0.1213	-0.0304	3.3	0.027
Na_2CO_3	101.325	373.150	1.3347	-0.5586	0.1619	2.0	0.010
Na_2SO_4	101.325	373.150	1.1918	-0.2076	0.04161	3.0	0.050
$Na_2S_2O_3$	101.325	373.150	0.9213	0.1446	-0.0081	9.0	0.032
$Na_2C_4H_4O_6$	101.325	373.150	1.1860	-0.0225	0.00194	7.0	0.008
$NaKC_4H_4O_6$	101.325	373.150	1.2552	-0.0048	0.00014	14.0	0.176
K_2CO_3	101.325	373.150	1.0362	-0.4025	0.4463	1.0	0.005
K_2SO_4	82.176	367.392	1.6327	-0.9273	0.3702	1.3	0.011
	101.325	373.150	1.1968	-0.2576	0.1340	1.8	0.063
	102.345	373.431	1.1149	-0.1049	0.0169	1.4	0.005
K_2CrO_4	101.325	373.150	1.0455	0.1065	-0.0188	1.6	0.028
$K_2C_2O_4$	101.325	373.150	0.9566	0.1983	-0.0324	4.2	0.039
$K_2C_4H_4O_6$	101.325	373.150	1.1930	0.0492	-0.0030	11.0	0.129
$MgCl_2$	99.992	372.779	0.9973	0.4317	0.0092	2.3	0.019
	101.325	373.150	1.4433	0.3851	0.0099	3.5	0.012
$CaCl_2$	101.325	373.150	1.7234	0.1326	0.0365	2.9	0.042
$CaBr_2$	101.325	373.150	0.5466	0.2695	-0.0147	5.3	0.107
$Ca(NO_3)_2$	101.325	373.150	1.7140	-0.0434	0.00651	8.7	0.022
$SrCl_2$	101.325	373.150	1.1066	0.3296	-0.0187	5.9	0.077
$SrBr_2$	101.325	373.150	1.6773	0.0976	-0.0007	10.6	0.429
$Sr(NO_3)_2$	98.819	372.450	2.1029	-0.3853	0.2515	2.1	0.031
	101.325	373.150	1.8513	0.1745	-0.0166	2.9	0.075
$BaCl_2$	73.319	364.368	0.7763	0.5734	-0.0855	2.5	0.034
	82.735	367.575	1.2163	0.1512	0.02687	2.5	0.043
	99.045	372.514	1.0730	0.3310	-0.0504	2.6	0.016
	100.365	372.884	1.0841	0.3560	-0.0579	2.6	0.026
	101.144	373.100	1.0805	0.2858	-0.0357	2.8	0.009
	101.325	373.150	1.5886	0.0657	-0.0103	2.7	0.051
$Ba(NO_3)_2$	82.740	367.576	1.1623	-0.3538	0.1203	1.2	0.007
	101.000	373.060	1.2633	-0.3713	0.1095	1.3	0.006
	101.325	373.150	1.7202	-2.0492	1.3379	1.1	0.001
$CdCl_2$	101.325	373.150	0.3631	-0.0288	0.00346	6.6	0.021
CdI_2	101.325	373.150	0.4205	0.2688	-0.0993	2.0	0.017
$HgCl_2$	101.325	373.150	0.4791	-0.0613	-0.0062	1.9	0.004
$Pb(C_2H_3O_2)_2$	101.352	373.150	0.3996	-0.0044	0.00007	19.9	0.022
$Pb(NO_3)_2$	98.819	372.450	1.0074	-0.1115	0.0189	2.8	0.016
	99.885	372.750	0.9701	-0.0712	0.0085	4.0	0.022
	101.325	373.150	1.1760	-0.1500	0.0172	4.0	0.036

Table 4. Coefficients of the Pitzer equations (16) for 1:2 and 2:1 electrolytes

	A_ϕ	$\beta^{(0)}$	$\beta^{(1)}$	C^ϕ	m_{max}/m^0	$\sigma(\phi)$	Ref.
$(NH_4)_2SO_4$	0.4501	0.0694	-9.6096	-0.0042	8.8	0.001	4
	0.4538	0.0585	-0.4298	-0.0032	8.1	0.001	4
	0.4596	0.0604	0.5865	-0.0035	6.2	0.005	4
	0.4603	0.0516	0.6013	-0.0025	8.7	0.002	6
	0.4606	0.0436	7.3960	-0.0016	7.5	0.002	4
Li_2SO_4	0.4603	0.1396	-1.2959	-0.0221	3.3	0.009	6
Na_2CO_3	0.4603	-0.0289	3.1000	0.0193	2.0	0.001	2
Na_2SO_4	0.4603	-0.6037	7.8147	0.3131	1.4	0.029	34
$Na_2S_2O_3$	0.4603	0.1302	0.9249	-0.0064	9.0	0.021	1
$Na_2C_4H_4O_6$	0.4603	0.0765	9.3471	-0.0040	7.0	0.004	1
$NaKC_4H_4O_6$	0.4603	0.0711	16.7445	-0.0026	14.0	0.028	1
K_2CO_3	0.4603	1.2595	-4.2757	-1.2145	0.4	0.001	30
K_2SO_4	0.4536	-0.1353	6.0693	0.0852	1.3	0.001	4
	0.4603	0.2669	0.7187	-0.0860	1.3	0.017	1,2
	0.4606	0.1187	1.5068	-0.0205	1.4	0.001	2
K_2CrO_4	0.4603	0.3117	-0.1641	-0.0769	1.6	0.010	2
$K_2C_2O_4$	0.4603	0.2018	-0.1007	-0.0223	4.2	0.012	29
$K_2C_4H_4O_6$	0.4603	0.0973	12.9378	-0.0041	11.0	0.015	29
$MgCl_2$	0.4599	0.3759	0.0552	-0.0274	2.3	0.008	5
	0.4603	0.4218	4.3736	-0.0247	3.5	0.011	1
$CaCl_2$	0.4603	0.3840	6.6072	-0.0273	2.9	0.007	1
$CaBr_2$	0.4603	0.1392	-3.6287	-0.0068	5.3	0.012	6
$Ca(NO_3)_2$	0.4603	0.1385	18.5582	-0.0061	8.7	0.021	1
$SrCl_2$	0.4603	0.3249	2.0443	-0.0267	4.8	0.025	1
$SrBr_2$	0.4603	0.2578	7.6829	-0.0121	10.6	0.087	6
$Sr(NO_3)_2$	0.4595	0.1952	10.3395	0.0489	2.1	0.010	4
	0.4603	0.4014	8.7646	-0.0401	2.9	0.003	4
$BaCl_2$	0.4502	0.3431	-0.1924	-0.0363	2.5	0.001	4
	0.4538	0.2650	2.8757	-0.0148	2.5	0.002	4
	0.4595	0.3793	0.2854	-0.0583	2.6	0.021	5
	0.4600	0.4117	0.1672	-0.0672	2.6	0.021	5
	0.4602	0.2644	1.9064	-0.0262	2.8	0.011	4
	0.4603	0.3161	5.5183	-0.0391	2.7	0.004	1
$Ba(NO_3)_2$	0.4538	0.0607	1.6641	-0.0076	1.2	0.001	4
	0.4602	0.0543	2.4292	-0.0033	1.3	0.001	4
	0.4603	-0.6504	6.3547	0.4103	1.1	0.001	1
$CdCl_2$	0.4603	-0.0491	-6.3452	0.0036	6.6	0.002	6
CdI_2	0.4603	0.1722	-5.0429	-0.0597	2.0	0.009	6
$HgCl_2$	0.4603	0.0671	-5.0901	-0.0423	1.9	0.015	5
$Pb(C_2H_3O_2)_2$	0.4603	-0.0363	-4.5483	0.0026	7.0	0.006	1
$Pb(NO_3)_2$	0.4595	0.0826	0.5850	-0.0127	2.8	0.005	4
	0.4598	0.0467	1.6399	-0.0037	4.0	0.001	4
	0.4603	0.0881	2.1165	-0.0113	4.0	0.013	1

Table 5. Boiling point elevations (equation (24)) for 2:2 electrolytes.

	P/kPa	T/*K*	*A*	*B*	*C*	m_{max}	*σ(θ)*
$MgSO_4$	101.325	373.150	0.4848	0.1502	-0.0035	6.2	0.075
$MnSO_4$	98.525	372.367	0.4811	-0.1653	0.0808	1.6	0.006
	101.325	373.150	0.3816	0.0410	-0.0018	4.5	0.017
$FeSO_4$	98.392	372.329	0.4864	-0.1712	0.0723	2.3	0.018
	101.325	373.150	0.4965	-0.1716	0.0726	2.3	0.018
$CoSO_4$	98.125	372.254	0.4007	-0.0120	0.0272	2.1	0.011
	101.325	373.150	0.4119	-0.0162	0.0295	2.1	0.011
$NiSO_4$	98.259	372.291	0.5561	-0.2157	0.1155	2.4	0.024
	101.325	373.150	0.5723	-0.2235	0.1191	2.4	0.026
$CuSO_4$	98.925	372.480	0.3322	0.0101	0.0210	4.6	0.025
	101.325	373.150	0.3332	0.0147	0.0207	4.6	0.025
$ZnSO_4$	98.992	372.499	0.2365	-0.0124	0.0410	2.8	0.034
	101.325	373.150	0.1818	0.1941	-0.0097	5.3	0.070
$CdSO_4$	98.832	372.454	0.3977	-0.0657	0.0350	2.6	0.022
	101.325	373.150	0.2985	0.0501	0.0059	2.6	0.030

Table 6. Coefficients of the Pitzer equations (17) for 2:2 electrolytes

	A_ϕ	$\beta^{(0)}$	$\beta^{(1)}$	$\beta^{(2)c}$	C^ϕ	m_{max}/m^0	$\sigma(\phi)$
$MgSO_4$[a,d]	0.4603	0.3476	7.2986	-37.23	-0.0168	6.2	0.028
$MnSO_4$[b]	0.4594	0.2143	3.7549	-40.00	0.0206	1.6	0.009
	0.4603	0.2729	3.8633	-40.00	-0.0228	4.5	0.020
$FeSO_4$[a,b]	0.4593	0.3521	2.7588	-40.00	-0.0519	2.3	0.021
	0.4603	0.2755	3.5141	-40.06	-0.0171	2.3	0.008
$CoSO_4$[b]	0.4592	0.1922	3.5096	-30.70	0.0461	2.1	0.036
	0.4603	0.1445	3.9053	-30.70	0.0719	2.1	0.043
$NiSO_4$[b]	0.4593	0.1685	4.8040	-40.06	0.0584	2.4	0.012
	0.4603	0.1345	5.1885	-40.06	0.0751	2.4	0.016
$CuSO_4$[b]	0.4595	0.2770	2.9135	-48.33	-0.0134	4.6	0.015
	0.4603	0.2914	2.8051	-48.07	-0.0163	4.6	0.019
$ZnSO_4$[b]	0.4595	0.3411	2.7509	-32.81	-0.0299	2.8	0.010
	0.4603	0.3512	2.7704	-32.81	-0.0265	5.3	0.029
$CdSO_4$[a]	0.4595	0.1684	4.0836	-48.07	0.0200	2.6	0.017
	0.4603	-0.0177	4.7595	-48.07	0.1100	2.6	0.080

[a] Ref. [1].
[b] Ref. [5]; [c] Ref. [24].
[d] Ref. [33].

CONCLUSION

The ebullioscopic method is one of several experimental techniques which are used to determine colligative properties of electrolyte solutions. The method is capable of yielding activities over a wide concentration range and by the change of external pressure, over a wide temperature interval. However, precise measurements of boiling temperatures are difficult to perform and, unfortunately, ebullioscopic determinations are less and less reported in the literature.

The majority of accessible boiling point data for electrolyte solutions, at various pressures, were critically assessed using third order polynomials of the boiling point elevations $\theta(m)$. The small values of the standard deviations $\sigma(\theta)$ clearly indicate an excellent agreement between the correlated values of the boiling point elevations and the experimental results.

The osmotic coefficients of the 1:1, 1:2, 2:1 and 2:2 type electrolytes were calculated from the correlated $\theta(m)$ values and represented in the framework of the Pitzer equations for the osmotic and activity coefficients. The temperature and pressure dependence of activity data as derived from the ebullioscopic determinations is of significant importance considering that there is increasing interest in the thermodynamic properties of electrolyte solutions at temperatures much above room temperature. No doubt that such data is useful in the theory of electrolytes, in chemical industry, desalination processes and geochemistry.

REFERENCES

[1] Gerlach, G.Th. *Z Anal.Chem.* 1887, 26, 413-531.

[2] Baroni, G. *Gazz. Chim. Ital.* 1893, *23*, 263.

[3] Schlamp, A. *Z. Physik. Chem. A*, 1894, *14*, 272.

[4] Buchanan, G.Y. *Trans. Roy. Soc. Edinb.* 1899, *29*, 529.

[5] Kahlenberg, L. *J. Phys. Chem.* 1901, *5*, 339-392.

[6] Johnston, S.M. *Trans. Roy. Soc. Edinb.* 1906, *45*, 193.

[7] Jablczynski, K.; Kon, S. *J. Chem. Soc.* 1923, *123*, 2953-2963.

[8] Swietoslawski, W.; Ebulliometric Measurements; Reinhold Publ. Co., New York, 1945.

[9] Saxton, B.; Smith, R.P. *J. Am. Chem. Soc.* 1932, *54*, 2626-2636.

[10] Smith, R.P. *J. Am. Chem. Soc.* 1939, *61*, 500-503.

[11] Smith, R.P.; Hirtle, D.S. *J. Am. Chem. Soc.* 1939, *61*, 1123-1126.

[12] Johnson, G.C.; Smith, R.P. *J. Am. Chem. Soc.* 1941, *63*, 1351-1353.

[13] Harned, H.S.; Owen, B.B.; *The Physical Chemistry of Electrolytic Solutions*; Third ed. Reinhold Publ. Co., New York, 1958.

[14] Robinson, R.A.; Stokes, R.H. *Electrolyte Solutions*; Butterworths, Sec. Ed., London, 1965.

[15] Ambrose, D. Vapour pressures. In *Experimental Thermodynamics*, Volume II., Le Neindre, B.: Vodar, B. Eds.; Butterworths, London, 1975; Vol. 2. pp. 607- 651.

[16] Desnoyers, J.E.; Experimental techniques: cryoscopy and other methods. In *Activity Coefficients in Electrolyte Solutions*; Pytkowicz, R.M.; Ed.; CRC Press, Inc. Boca Raton, 1979; Vol. 1, pp. 139-155.

[17] Timmermans, J. *The Physico-chemical Constants of Binary Systems in Concentrated Solutions. Systems with Metallic Compounds*. Interscience Publ.Inc. New York, 1960; Vol.3.

[18] Timmermans, J. *The Physico-chemical Constants of Binary Systems in Concentrated Solutios. Systems Inorganic+Organic or Inorganic Compounds (Excepting Metallic Derivatives)*. Interscience Publ. Inc. New York, 1960; Vol.4.

[19] Washburn, E.W., Ed.; *International Critical Tables and Numerical Data: Physics, Chemistry, and Technology*. McGrow-Hill, New York; 1926-1933; Vol. 1-7.

[20] Lansdolt-Börnstein - *Zahlenwerte und Funktionen aus Physik, Chemie, Astronomie, Geophysik, und Technik.*Sixth Ed., Springer-Verlag, Berlin; 1950.

[21] Horvath, A.L. *Handbook of Aqueous Electrolyte Solutions. Physical Properties, Estimation and Correlation Methods*; Ellis Horwood Ltd, New York, 1985.

[22] Apelblat, A. *AIChE Journal* 1993, *39*, 918-923.

[23] Meranda, D.; Furter, W.F. *J. Chem. Eng. Data* 1977, *22*, 315-317.

[24] Pitzer, K.S., Theory: Ion interaction approach. In *Activity Coefficients in Electrolyte Solutions*; Pytkowicz, R.M.; Ed.; CRC Press, Inc. Boca Raton, 1979; Vol. 1, pp. 157-208.

[25] Saul, S.; Wagner, W. *J. Phys. Chem. Ref. Data* 1987, *16*, 893.

[26] Dymond, J.H.; Smith, E.B. The Virial Coefficients of Pure Gases and Their Mixtures. A Critical Compilation. Clarendon Press, Oxford, 1980.

[27] Plake, E. *Z. Physik. Chem.* A. 1935, 172, 113-128.

[28] Rosenheim, A.; Liebknecht, O. *Ann. Chem.* 1899, *308*, 40.

[29] Koppel, J.; Cahn, M. *Z. Anorg. Chem.* 1908, *60*, 53.

[30] Jones, H.C.; Getman, F.H. *J. Am. Chem. Soc.* 1904, *31*, 303.

[31] Fredenhagen, K.; Wellmann, M. *Z. Physik. Chem.* A. 1932, 162, 454-467.

[32] Fredenhagen, K.; Kerck.E. *Z. Anorg. Chem.* A. 1944, *252*, 252-280.

[33] Gerlach, H. *Caliche* 1926, *8*, 298.

[34] Bancroft, W.D.; Gould, L.P. *J. Phys. Chem.* 1934, *38*, 197-211.

[35] Dietz, A.A.; Degering, E.F.; Schopmeyer, H.H. *Ind. Eng. Chem.* 1941, *33*, 1444.

[36] Campbell, A.N.; Kartzmark, E.M. *Can. J. Res.* 1950, *28*, 43.

[37] Roloff, M. *Z. Physik. Chem.* A. 1893, 11, 7.

[38] Vercher, E.; Vazquez, M.I.; Martinez-Andreu, A. *J. Chem. Eng. Data* 2001, *46*, 1584-1588.

[39] Vercher, E.; Vazquez, M.I.; Martinez-Andreu, A. *Fluid Phase Equilib.* 2002, *202*, 121-132.

[40] Vercher, E.; Vazquez, M.I.; Martinez-Andreu, A. *J. Chem.Eng. Data* 2003, *49*, 566-569.

[41] Miles, M.H.; Arman, H.D.; Carrick, J.D.; Gren, C.K.; Haggerty, K.A.; Kim, H.Y.; Ky, A.G.; Markham, J.E.; Meeks, C.F.; Noga, D.E.. *The Ninth International Conference on Cold Fusion*. Beijing, China, Tsinghua University, 2002.

In: Solution Chemistry Research Progress
Editor: Darian V. Bostrelli, pp. 149-161
ISBN 978-1-60021-993-1

Chapter 7

SOLUBILITY ISOTHERMS AS SOURCES OF STRUCTURAL INFORMATION ABOUT SATURATED SOLUTIONS

Jitka Eysseltova
Charles University, Prague, Czech Republic

ABSTRACT

Several ways of obtaining information in ionic processes by the analysis of solubility isotherms have been demonstrated. The computational treatment called «hydration analysis » has been described. The special properties of saturated solutions of hydrates have been demonstrated an discussed. Finally, the information obtainable by the comparative analyses of solubility isotherms in systems with an electrolyte, an organic solvbent and water has been discussed and a formalism to describe the shape of such isotherms has been suggested.

I. HYDRATION ANALYSIS AS A WAY OF THE ANALYSIS OF SOLUBILITY ISOTHERMS

Solubility is a very complex property, reflecting a lot of energetic contributions as hydration energies of the ions involved, energy of ion pairing and complexing in saturated solutions, crystallization energies of the equilibrium solid phases etc. Its sensibility to the ionic processes taking place in saturated solutions is especially striking in the case of ternary systems containing pairs of salts with a common ion and water. As an example, let us examine the behavior of $LiNO_3$ 3 H_2O when different bivalent ions are added. According to our own measurements [1] eutonic systems exist at 25 °C between $LiNO_3$ 3 H_2O and $Mg(NO_3)_2$ 6 H_2O, $Sr(NO_3)_2$ and $Ba(NO_3)_2$. On the other hand, $LiNO_3$ 3 H_2O dehydrates under the influence of $Ca(NO_3)_2$ 4 H_2O, anhydrous $LiNO_3$ existing in contact with the saturated solution in broad concentration range [2]. Finally, series of dehydration of both components occurs [3] when $Co(NO_3)_2$ 4 H_2O is added to saturated solution of $LiNO_3$ 3 H_2O under identical conditions. The respective isotherms are depicted in molality scale at Figure 1 so that the differences among the isotherms are demonstrated clearly. It is to be seen, that the systems under consideration differ not only as far the quality of occurring solid phase discussed above is concerned, but also as far the position of invariant points on the isotherms, which can be understood as the length and or mutual position of the solubility branches, is concerned. Shortly, the addition of Mg^{2+} and Co^{2+} to saturated solutions of $LiNO_3$ 3 H_2O has similar effect at the solubility of the later, reducing its concentration in saturated solution (salting out). In the case of Mg^{2+}, this reduction goes on to cca 10 mol kg^{-3} when $Mg(NO_3)_2$ 6 H_2O crystallizes. In the case of $LiNO_3$ 3 H_2O, the concentration of $LiNO_3$ 3 H_2O starts to rise again and the next solid phase the lower hydrate $LiNO_3$ 2 H_2O starts to crystallize. Similarily, first addition of both Ca^{2+} and Sr^{2+} to saturated solutions of $LiNO_3$ 3 H_2O let the concentration of the later unchanged. In the case of Sr^{2+}, anhydrous $Sr(NO_3)_2$ being the neighbouring solid phase at about 2 mol Sr^{2+} pro 1000 g H_2O. At higher concentration of Ca^{2+} a large area of retrograde solubility with massive rise of the $LiNO_3$ 3 H_2O concentration (salting in) occurs, $LiNO_3$ 3 H_2O being finally dehydrated to anhydrous $LiNO_3$. The differences just described cannot be put in relation to any variables such as solubility or degree of hydration of the added components or radius of the added ion. Therefore ionic processes taking place in the saturated solution are the most probable reason and getting information about these processes by the analysis of solubility isotherms is very tempting, as a lot of material about solubility has been collected, e.g. in the materials of the Solubility Data Project of IUPAC [4].

In recent years [1,5–7], we have tried to search for a method of computer analysis of solubility curves that would elucidate this information. The proposed method called hydration analysis [5], permits the calculation and, for the required treatment, use of the amount of water that attains different properties in the three component saturated solution from those in the saturated binary solution.

In hydration analysis, the amount of water which preserved in ternary solution the properties of water in binary saturated solution of the uni-univalent electrolyte MX (the effective water w_{eff}) is calculated so that

$$m_0^2 = m_{\mathrm{MX}}\,(m_{\mathrm{MX}} + m_{\mathrm{M'X}})\,. \tag{1}$$

This may be done by expressing the molalities via mass %:

$$\left(\frac{1000\,w_0}{(M_r)_{\mathrm{MX}}\,(w_{\mathrm{H_2O}})_0}\right)^2 = \frac{1000\,w_{\mathrm{MX}}}{(M_r)_{\mathrm{MX}}\,w_{\mathrm{eff}}}\;\frac{1000}{w_{\mathrm{eff}}}\left(\frac{w_{\mathrm{MX}}}{(M_r)_{\mathrm{MX}}} + \frac{w_{\mathrm{M'X}}}{(M_r)_{\mathrm{M'X}}}\right), \tag{2}$$

where w_0 is the concentration in mass % (or molality m_0) of the binary saturated solution serving as our reference point and $(w_{\mathrm{H_2O}})_0$ is the concentration of water in that solution. From (2),

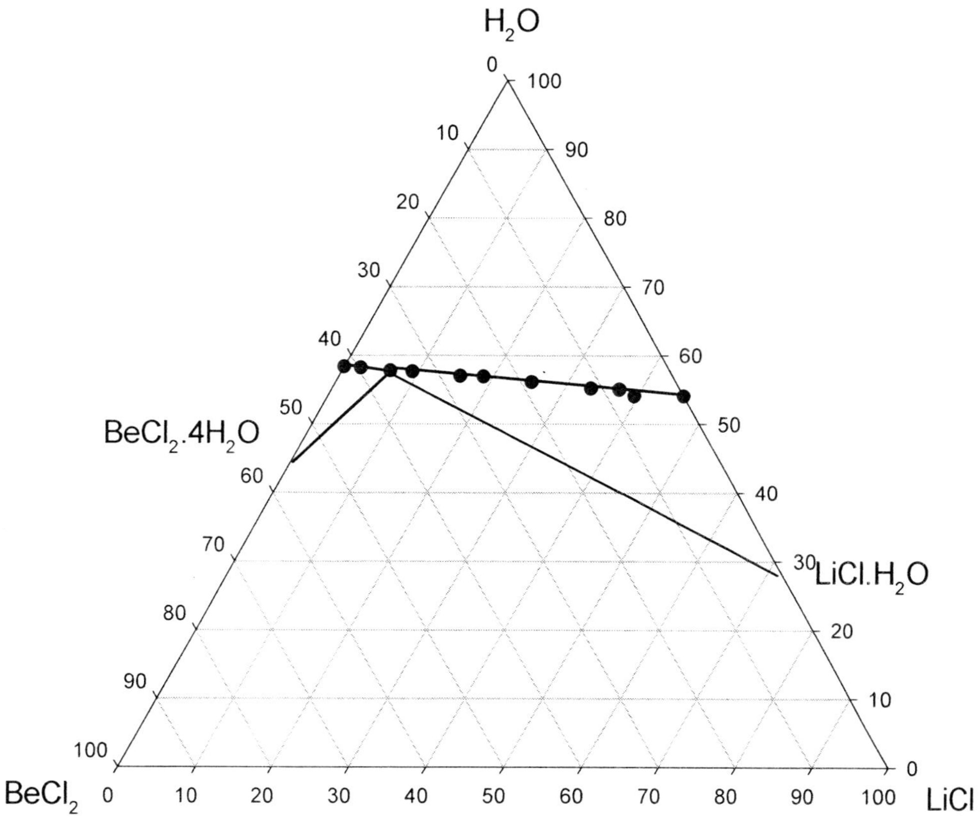

Figure 1. Solubility in the $BeCl_2 - LiCl - H_2O$ system at 298 K.

$$w_{\text{eff}} = \frac{(w_{H_2O})_0\,(M_r)_{MX}}{w_0}\sqrt{n_{MX}\,(n_{MX} + n_{M'X})}\ , \tag{3}$$

where $n_i = w_i / (M_r)_i$ is the amount of substance of the component i in 100 g of the saturated solution. Therefore, the amount of water that attains different properties in the three component saturated solution from those in the saturated binary solution $w_{\text{changed}} = w_{H_2O} - w_{\text{eff}}$ is expressed as the parameter P:

$$P = \frac{w_{H_2O} - w_{\text{eff}}}{(M_r)_{H_2O}\,(n_{H_2O} + n_{MX} + n_{M'X})}, \tag{4}$$

where $w_{H_2O} = 100 - w_{MX} - w_{M'X}$ answers to the concentration of water in ternary saturated solution under consideration (in mass %). For the sake of clarity, the parameter P is expressed relative to one mole of the non-saturating component, i. e. P/x_2.

From the values obtained by hydration analysis, the concentration dependence of the relation P/x_2 and the total amount of water available per ion of salt components Aq were found as the most suitable for the description of ionic processes [5].

It cannot be expected that an isolated set of values of the parameters obtained by the hydration analysis on one solubility branch will have big information value. The opposite is true, if the hydration analysis is applied to families of homologous systems. Such comparative analysis, involved and widely used in electrolyte chemistry by Lilich and his school [8,9] can lead to obtaining valuable additional information based on solubility data available in the literature, as demonstrated on model systems [6, 10].

Among this additional information, also some opportunity to transpose the concept of hydration analysis to the terms more understandable for thermodynamics and physical chemistry in general exists. It is straightforward [6] to obtain the ratio $\gamma_{\pm}/\gamma_0$ ($\gamma_{\pm}$ and γ_0 are respectively activity coefficients of the solute in the ternary saturated solution and in its binary saturated solution in water under the same conditions), as

$$P = \frac{w_{H_2O}}{(M_r)_{H_2O}\,(n_{H_2O} + n_{MX} + n_{M'X})}\left(1 - \frac{\gamma_0}{\gamma_{\pm}}\right) = x_{H_2O}\left(1 - \frac{\gamma_0}{\gamma_{\pm}}\right), \tag{5}$$

where x_{H_2O} is the mole fraction of water in the ternary saturated solution under consideration.

In the connection with the shape of the equation (5), the attention should be turned to the fact that in molality scale, the free energy of transfer of the solute 1 from its binary saturated solution to a ternary saturated solution is defined by the equation

$$\Delta G_t = v_1 \, RT \ln(\gamma_1 m_1 / \gamma_{10} m_{10}) \tag{6}$$

where v_1 is the number of moles of ions formed from 1 mole of solute 1, R is the gas constant, T the temperature in K, m_{10} and γ_{10} are the molality and and the activity coefficient of solute 1 in its binary saturated solution respectively and m_1 and γ_1 are the corresponding quantities in a ternary solution.

The Eq. (6) can be rearranged to

$$\Delta G_t = \Delta G_I + \Delta G_{II} \tag{7}$$

where

$$\Delta G_I = v_1 \, RT \ln(\gamma_1 / \gamma_{10}) \tag{8}$$

and

$$\Delta G_{II} = v_1 \, RT \ln(m_1 / m_{10}) \tag{9}$$

In Eqs. (7) and (8), the ΔG_I means the change in excess free energy (or non-ideal free energy), the ΔG_{II} is the change in ideal free energy during the respective transfer. The transfer of electrolytes from water into organic liquids or mixtures of organic liquids with water has been studied widely; on the other hand, only limited amount of material on transfer of a solute from a binary to a ternary water solution may be localized (e.g.[11–14]). Even in these cases, the component B in the ternary electrolyte – B – H_2O system is an organic non-electrolyte. As far the author is aware, no information about thermodynamics of transfer in systems consisting of two electrolytes and water has been published as yet in spite of the fact that the relation of activity coefficients such as that in the equation (5) are available easily. In connection with these ideas, the quantity Ξ has been introduced [7] as

$$\Xi = m_1 \mathrm{R}T \ln \gamma_\pm / \gamma_0. \quad (10)$$

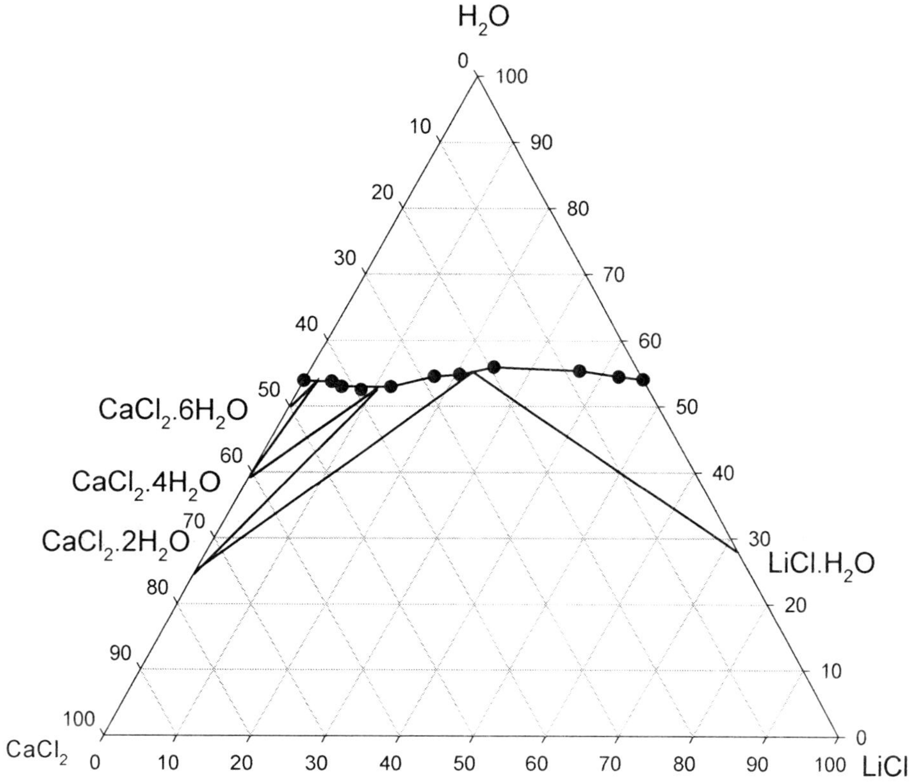

Figure 2. Solubility in the $CaCl_2$ – LiCl – H_2O system at 298 K.

Without detailed considerations of the pertinent standard states, this quantity can be assigned the significance of the residual Gibbs energy for the transfer of m_1 moles of electrolyte 1 from its saturated solution to a solution containing 1 kg of water and a concentration of electrolyte 2 such that the ternary solution formed at temperature T is exactly saturated in electrolyte 1.

It has been proven [7], that the dependence of the quantity Ξ on the composition along the branches of the solubility isotherm of ternary system follows the changes of configuration of the ternary saturated solution.

II. Specific Features of the Saturated Solutions of Hydrates

During selecting of interesting families of systems, both by literature search and own measurement, special features of solubility isotherms of salt hydrates have been observed. The 25 °C solubility isotherms in the systems $MnCl_2$ – $MgCl_2$ – H_2O [15], $CoCl_2$ – $MgCl_2$ – H_2O [16] and $CuCl_2$ – $MgCl_2$ – H_2O [17] were analyzed by the above described hydration analysis. When magnesium chloride is added to saturated solutions of $CoCl_2$ 6 H_2O, $MnCl_2$ 4 H_2O and $CuCl_2$ 2 H_2O, hydration of the Mg^{2+} ion added by the water molecules available in the saturated solution proceeds. The parameter P/x_2 giving quantitative characterization of the competition is approximately 3, 4, and 5.5, respectively. This sequence ($CoCl_2$ 6 H_2O < $MnCl_2$ 4 H_2O < $CuCl_2$ 2 H_2O) does not reflect the amount of water available in the respective saturated solutions which increases in the order $MnCl_2$ 4 H_2O < $CuCl_2$ 2 H_2O < $CoCl_2$ 6 H_2O. Instead, parameter P/x_2 seems to be the larger, the lower the water content in the hydrate at equilibrium. This means: the higher hydrate is present in the equilibrium solid phase, the less willing the respective saturated solution is to let the Mg^{2+} ion build its hydration sphere.

In the systems with hydrated salts, quite often the solubility branches of the hydrated components are too short to meet the criteria established for applicability of hydration analysis [18]. The phenomenon itself can be put into connection with the problem under discussion as well. Let us assume that a saturated solution of a hydrate has its own configuration, more or less related to the structure of the respective crystal hydrate. Such relation between concentrated electrolyte solutions and solid crystal hydrate was assumed earlier in some considerations of Mishchenko [19] and reflected also by using ionic radii based on solid hydrate data in attempts to use Born equation for consideration of ion hydration [20]. Such configuration represents a minimum of potential energy of the system. Adding the other solute causes a perturbation of this state and hence a potential energy increase. In the above systems with $MgCl_2$ added to the liquid phase, the added component has the ability to organize the solution as well due to strong hydration of magnesium cation. Therefore, the respective hydration energy contributes to the compensation of breaking the original configuration. In most other cases, the system maintains its potential energy by expelling the added component into the solid phase, regardless of whether in the pure form (forming an eutonic system of unsymmetrical shape) or as a double salt (often e.g. in the case of $MgCl_2$ 6 H_2O in systems with alkali metal chlorides). Such phenomenon is reflected in the shape of the solubility isotherm under consideration as a low concentration of the non-saturating component in the respective eutonic point, or, in other words, as a very short crystallization branch of the hydrate.

On the other hand, it should be stressed that it is not only the ion hydration in the saturated solution, but the overall configuration of the respective solution, involving packing of its ions and water molecules in space, which is responsible for the phenomena under consideration. This may be illustrated by the following example: as mentioned above, of chloride systems, $MgCl_2$ 6 H_2O is the hydrate, which exhibits very often a short crystallization branch due to organization of its saturated solution. However, in the $CaBr_2$ – $MgBr_2$ – H_2O system at 25 °C [21], the solubility isotherm is highly unsymmetrical, with the eutonic point at 52.6 mass % $MgBr_2$ 6 H_2O and 6.7 mass % $CaBr_2$ 6 H_2O, in spite of the fact that solubility of both salts does not differ much ($MgBr_2$ 6 H_2O 50.8 mass %, i.e. 5.62 mol

kg^{-1}, $CaBr_2$ 6 H_2O 60.5 mass %, i.e. 7.66 mol kg^{-1}). This means that, in this case, $CaBr_2$ 6 H_2O is the component which exhibits the more organized saturated solution, expelling the added $MgBr_2$ into solid phase. This fact may be accepted and understood if we take into account the size of the ions involved ($r_{Mg^{2+}}$ = 65 pm, $r_{Ca^{2+}}$ = 99 pm, r_{Cl^-} = 181 pm and r_{Br^-} = 196 pm). From this point of view, the dimensions of Mg^{2+} and Br^- ions differ too much to ensure a favorable filling of the space, the situation being more favorable in the case of calcium.

The respective solubility isotherm was treated by the hydration analysis; on the solubility branch depicting the solubility of $MgBr_2$ 6 H_2O in $CaBr_2$ solutions, additional support for the above conclusions was found. In Figure 2, the dependences of the amount of available water Aq, the hydration parameter P/x_2 and the variable Ξ on the concentration of the added Ca^{2+} ion are presented. As we can see, the ions of the added electrolyte build their hydration spheres from the water available in a saturated solution of the solute. Moreover, a structure change characterized by a break on the concentration dependence of Ξ is demonstrated. This may be understood as adaptation of the structure to that more related to the energetically advantageous structure of the saturated solution of $CaBr_2$ 6 H_2O as the calcium concentration in the solution grows. On the shorter solubility branch depicting the solubility of $CaBr_2$ 6 H_2O in $MgBr_2$ solutions, the added magnesium bromide is smoothly incorporated into the structure of the binary saturated solution and expelled into the solid phase soon.

To explain this concept, let us analyze the addition of an other component to highly organized saturated solution. Two possibilities occur: first, the added component has the ability to organize the solution strongly as well. In such case the added component is accepted and a solubility branch occurs. Second and more frequented case is, that the added component is not able to contribute to the organization of the solution. In such case, the added component is expelled into the solid phase, eutonic solution containing only small concentration of the added component is formed and the crystalization branch is very short.

A more detailed insight into this problem may be got with the aid of adapted schemes frequently used for deriving binary phase diagrams from the dependencies of Gibbs free energy on composition. For the purpose of the problem under consideration, that approach should be adapted to isohydral sections through isotherm of such system. Such replacement of temperature by concentration variable is definitely possible and used frequently in physico-chemical analysis. We can imagine that the shape of the dependence of Gibbs free energy of liquid mixtures (solutions) on composition will have a shape similar to that in binary systems. However, instead of a line connecting the Gibbs energy of solid phases used in treatment of binary eutonic systems or a curve describing the dependence of Gibbs free energy of solid solutions, a set of curves will occur in our case, depicting the concentrational dependence of the Gibbs free energy in mixtures of equilibrium solid and liquid along the respective isohydra. In the case of a very narrow crystallization field a shape of the respective dependence of ΔG on composition must exhibit a sharp minimum. What can be responsible for such strong modification of the shape of this curve? Generally, we can write

$$\Delta G_{mix} = x_A \Delta H_A + x_B \Delta H_B + RT(x_A \ln x_A + x_B \ln x_B) - T x_A \Delta S_A - T x_B \Delta S_B \quad (11)$$

In the enthalpy terms, no special changes can be expected. Therefore it is the entropy therm which with greatest probability is responsible for the sharp minimum at this curve and for the narrow crystallization branch of the respective solid as well. This statement is in full correspondence with the idea of highly structured solutions stated a few minutes ago.

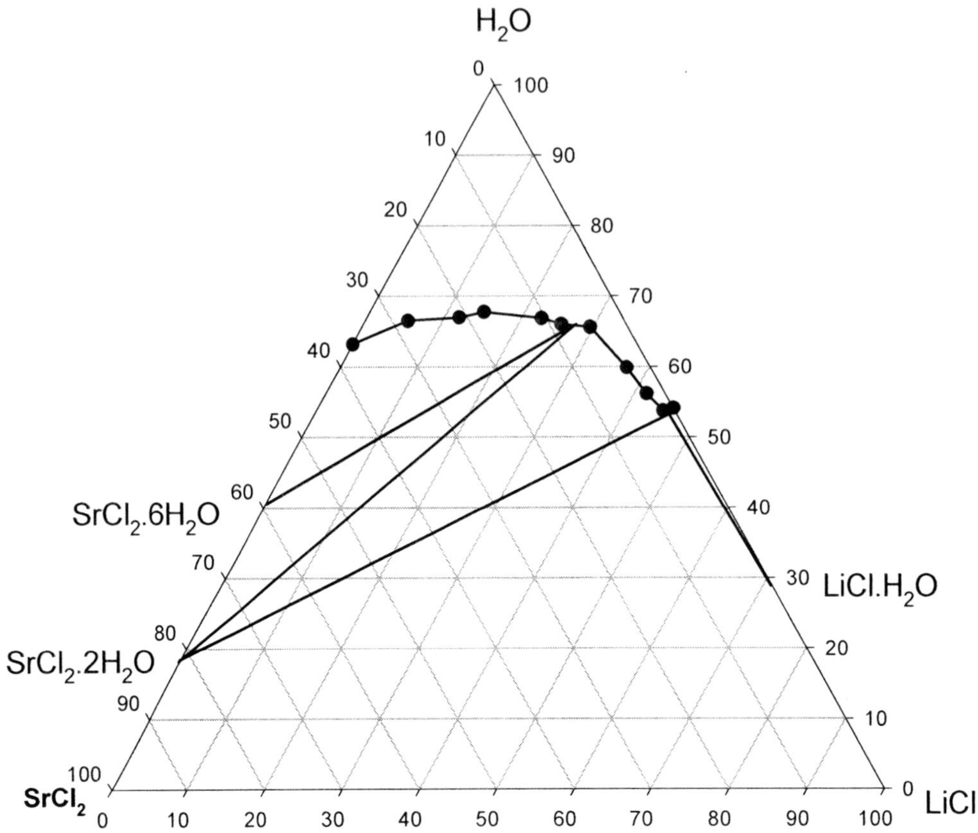

Figure 3. Solubility in the $SrCl_2$ – LiCl – H_2O system at 298 K.

Several other cases for illustration belong to chloride family [22,23]. In the $BeCl_2$ – LiCl – H_2O system (Figure 1), beryllium chloride tetrahydrate is the more organized component and in the same time beryllium is the component which builds its hydration sphere most intensively, even on account of the water molecules belonging originally to the dihydrate of LiCl. In the $MgCl_2$ – LiCl – H_2O system [23], the situation is difficult to discuss because crystallization of a compound must bring some other terms into the equation for Gibbs energies. The highly structured nature of the solution of lithium chloride monohydrate finds its expression together with the tendency of lithium ion to build its hydration sphere in the systems with the rest of alkaline earth chlorides. As the result, lower hydrates of calcium and strontium chlorides crystallize from the solution (Figures 2 and 3 respectively). The high organization of the saturated solution of lithium chloride monohydrate especially when compared with that of strontium chloride hydrates leads to shortening of its crystallization field. Finally, this effect is demonstrated most clearly in the system with barium chloride (Figure 4).

These ideas can on side enable us to construct series on which the organization of saturated solutions rises, which can be of valuable help for understanding the phenomena taking place in concentrated electrolyte solutions. on the other side, these ideas may have also direct application in designing of crystallization processes. It is clear, that successful purification of compounds yielding highly organized saturated solutions through crystallization multicomponent solutions – e.g. crystallization of magnesium salts from sea water – is difficult to suppose.

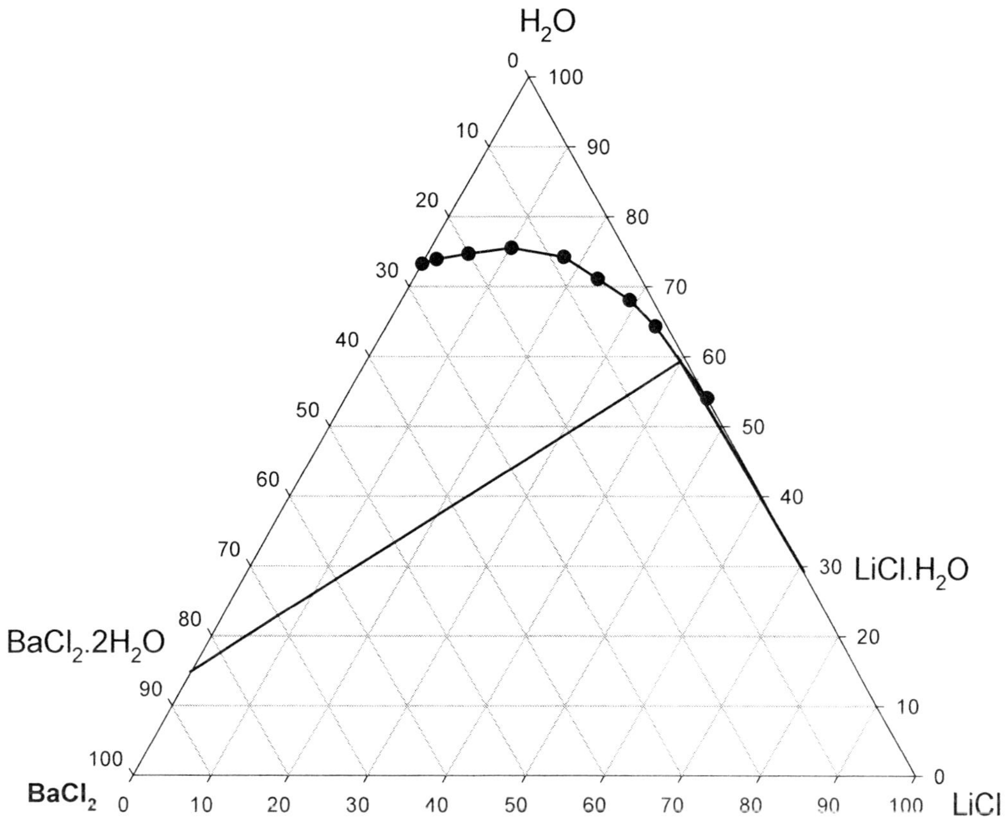

Figure 4. Solubility in the $BaCl_2$ – LiCl – H_2O system at 298 K.

III. Solubility in Systems Consisting of an Electrolyte, Water and Organic Solvent

The importance of the investigation of solubility of electrolytes in mixture of water and an organic liquid is obvious. However, the field suffers by shortage of systematic data, the interest being focused on the thermodynamics of transfer of sparingly soluble salts from water to mixed solvents [24]. On the other hand, investigation of systems with electrolytes exhibiting middle or even high solubility and even forming hydrates may bring new information relevant to the subject of this article.

A relatively limited number of systems of this kind are described as complete solubility isotherms in the literature. As a rule, isolated systems have been found and some families of systems varying either the organic solvent or an ion in the electrolyte are a real exception – e.g. [25–27]. Further on, the shape of the isotherms found in the literature is extremely heterogeneous. The solubility isotherms in systems differring only by some detail in the structure of the solvent or in one ion of the electrolyte can exhibit substantial differences. E.g., quite frequently in systems with potassium salts smooth solubility curve connecting the two binary solutions with more ore less salting out exists, whereas with sodium salts immiscibility occurs. In such case two invariant points belonging to the equilibrium of one solid and two liquid phases appear on the solubility curve, the phase diagram being therefore divided into five fields, which are to be seen on Figure 5, obtained by our own measurements in the system NaCl – dioxane – water [28].

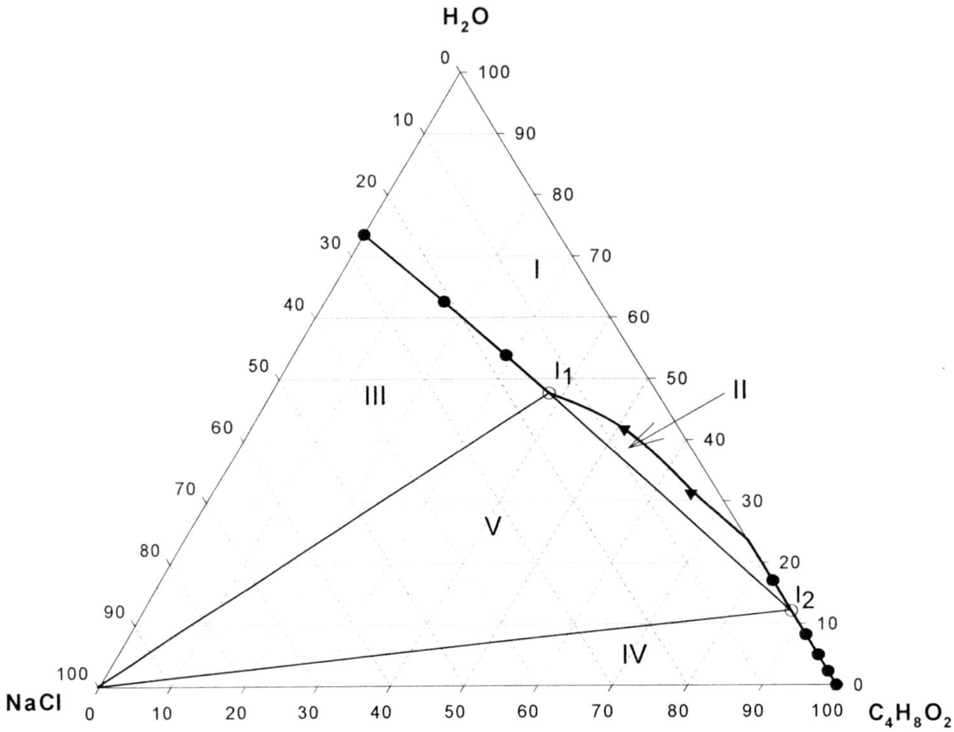

Figure 5. Solubility in the NaCl – dioxane – H_2O system at 298 K.

Field I belongs to unsaturated solutions, field II to the equilibria between two conjugated liquid phases, fields III and IV depict the equilibria between solid electrolyte and its saturated solution, water rich and organic solvent rich respectively, and field V depicts the equilibrium between solid electrolyte and the two invariant solutions I1 and I2.

If the electrolyte exists like a hydrate in the binary system electrolyte – water at 298 K, such hydrate must lose its water under the influence of the organic component and yield the anhydrous salt or even some adduct with the organic component being in equilibrium with the liquid phase in the binary system electrolyte – organic component. Such dehydration (occurring frequently in water systems caused by addition of acids to their salts) may be stepwise or in one step and again, slight change in the nature of components can change the shape of the isotherms very dramatically.

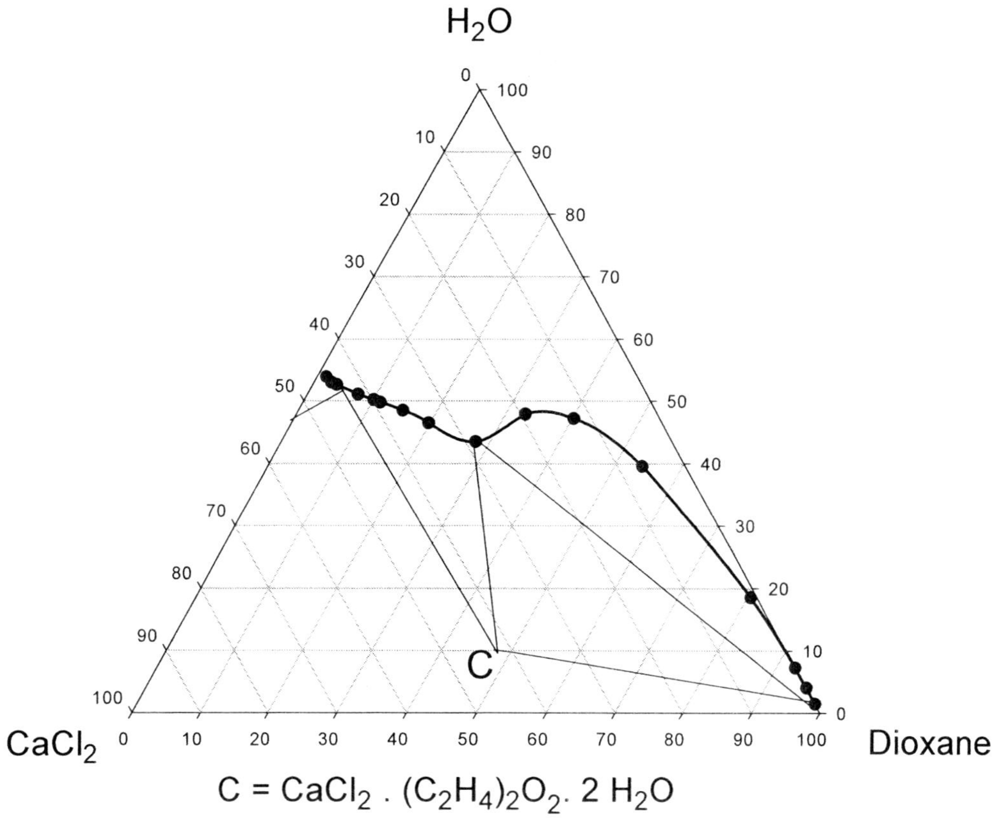

Figure 6. Solubility in the $CaCl_2$ – dioxane – H_2O system at 298 K.

The most complicated variation of this theme is the formation of a ternary compound, best understandable as hydrated adduct of the electrolyte and the organic component as an intermediate step of such dehydration process. In addition, the above discussed lose of miscibility can occure and occurs indeed also in just discussed systems with more solid phases. As a very interesting example the system $CaCl_2$ – dioxane – water [29] may serve (Figure 6).

Table 1. Suggestion of short description of the crystallization fields of isotherms in electrolyte – organic solvent – water systems

Anhydrous salt ... 0/0
n-hydrate ... *n*/0
Adduct with *n* organic molecules ... 0/*n*
n-hydrated adduct with m organic molecules ... *n*/*m*
formation of layers ... underline

The term of "congruent solubility" connected with the maximum on the respective solubility isotherms seems to lose its usual meaning in these cases. Identical shape of the solubility isotherm has been observed in our laboratory in the system LiCl – dioxane – water, differing sharply from the systems with sodium and potassium chlorides which behave in the way described above. The only statement which could be made about it is, that such field of coexistence of two liquid phases must be situated on only one crystallization field and cannot be spread over two – this is the consequence of the Phase Rule. Otherwise its existence or non-existence depends only on the components of the respective system reflecting therefore the ionic processes taking place in it (with the highest probability mainly the competitive solvation) quite dramatically.

Obviously, the problem is quite complex. To be able to discuss the material in some easy and lucid way, the formalism presented in Table 1 can be suggested. Each branch of the solubility isotherm has been described with the aid of composition of the equilibrium solid phase by a numerical code, which includes the number of water molecules in the hydrate being in equilibrium with water as the numerator and the number of molecules of organic solvent bound in the solvate being in equilibrium with organic solvent as the denominator. Thus, the system formed by one crystallization field of an anhydrous salt is described by simple code 0/0. If immiscibility occurs at the respective branch, the code is underlined. Therefore, the difference between the systems containing sodium and potassium salts water and dioxane may be described as "as a rule, the systems with sodium salts are of the type $\underline{0/0}$ whereas those with potassium salts are of the type 0/0". The most complicated system discussed here and depicted at Figure 6 may be described as 6/0 – $\underline{1/2}$ – 0/0.

Systematic investigation of new families of such systems varying either the ions involved or the organic component should be highly encouraged. On one hand new views into competitive salvations may be expected, on the other hand also the idea of the thermodynamic of transfer of the electrolyte from neat organic solvent to the binary solvents (where most electrolytes become sparingly soluble) could be an interesting complement to the investigations made in water rich solutions.

References

[1] Zbranek V., Eysseltová J. *Monatsh. Chem.* 2001. *132*,1463-9.

[2] Khripun M. K., Chervonenko K. Yu., Efimov A.Yu., Khripun V. D., Kiselev A. A. *Zh. Obshch. Khim.* 2000.*70/2*, 217 –22.

[3] S.Savenko, T. A. Andreeva, *Zh. Neorg. Khim.* 1975. *20*, 2514-9.

[4] R. Cohen-Adad, J.W. Lorimer, S.L. Phillips, M. Salomon, *J. Chem. Info. Comp. Sci* 1995. *35 (4)*, 675-87.

[5] J. Eysseltova, *Collect. Czech. Chem. Commun*1994. *59,* 126-30.

[6] J. Eysseltova, *Collect. Czech. Chem. Commun.* 1994. *59*, 2351-7.

[7] J. Eysseltova, M. Ebert, *Z. Naturforsch.* 1999. *54a*, 485-7.

[8] L. S. Lilich *Nekotorie aspekty sovremennogo predstavleniya o rastvorach elektrolitov (Some aspects of the modern knowledge of electrolyte solutions)* Leningrad university Publishers, Leningrad 1967.

[9] L. S. Lilich, S. A. Shchukarev *Zhurn.Neorg. Khim.* 1984 *29*, 347-56.

[10] J. Eysseltová, V. Zbranek, J. Jirsák: *Collect. Czech. Chem. Commun.* 1999. *64*,1262 -7.

[11] Jianji Wang, Wenbin Liu, Tongchuan Bai, Jionsu Lu J. *Chem. Soc. Faraday Trans.* 1993. *89(11),* 1741-4.

[12] Hasudo Uedaira *Bull. Chem. Soc. Japan* 1975. *48(7),* 2006-9.

[13] J. P. Chatterjee, I. N. Basamullick *J. Chem. Soc. Faraday Trans.* 1988. *84 (1)*, 2697-3001.

[14] K. Gekko *J. Biochem.* 1981. *90*,1633-6.

[15] I. Zhelnin, G. I. Gorshtein *Zh. Neorg. Khim.* 1971. *16*, 3143 -9.

[16] T. Oykova *Zh. Neorg. Khim.* 1979. *24*, 219-24.

[17] Kh. Balarev, D. Spasov *Zh. Neorg. Khim.* 1980. *25*, 2814-9.

[18] J. Eysseltova, *Collect. Czech. Chem. Commun.* 2001. *66*, 89-95.

[19] K. P. Mishchenko, G. M. Poltoratskiy *Voprosy Termodinamiki i Stroeniya Vodnykh i Nevodnykh Rastvorov Elektrolitov (Problems of Thermodynamics and Structure of Electrolyte Solutions)*, Khimiya, Leningrad 1968.

[20] N. E. Khomutov *Zh. Fiz. Khim.* 1965. *39*, 639-43.

[21] Kh. Balarev, R. Duhlev *Zh. Neorg. Khim.* 1983. *28*, 2639-46.

[22] V. P. Blidin, V. I. Gordyenko *Dokl. Akad. Nauk SSSR* 1954. *24*, 1081-9.

[23] N. K. Voskresenskaya, O. K. Yanat'eva *Izv. AN SSSR, ser. khim.* 1937. *1*, 97-102.

[24] Kalidas, G. Hefter,Y. Marcus, *Chem. Rev.* 2000. *100*, 820-52.

[25] V. Gomis, F. Ruiz, J.C. Asensi, M. D. Saquete. *J. Chem. Eng. Data* 1996. *41*, 188-91.

[26] V. Gomis, F. Ruiz, G. De Vera, E. Lopez, M. D. Saquete *Fluid Phase Equilib.* 1994. *98*, 141-7.

[27] R. De Santis, L. Manelli, P. N. Muscetta *Chem. Eng. J.* 1976. *11a*, 207-11.

[28] J. Eysseltová, Z. Málková *J. Sol. Chem.* 2006. *35*, 1329-34.

[29] H. F.Bogardus, C. C. Lynch, *J. Phys.Chem.* 1943. *47*, 650-4.

In: Solution Chemistry Research Progress
Editor: Darian V. Bostrelli, pp. 163-176
ISBN 978-1-60021-993-1

Chapter 8

INTERACTIONS OF SACCHARIDES WITH L-SERINE: VOLUMETRY, VISCOSITY AND INFRARED SPECTROSCOPY

Kelei Zhuo,* Qian Liu, and Jianji Wang

School of Chemistry and Environmental Science, Henan Normal University, Xinxiang, Henan 453007, P. R. China

ABSTRACT

Apparent molar volumes, $V_{\Phi,S}$, and viscosity *B*-coefficients for D-(+)-glucose, D-(+)-galactose, D-(+)-xylose, and D-(-)-ribose in aqueous L-serine solutions have been determined respectively from density and viscosity measurements at 298.15 K. The infinite-dilution apparent molar volumes of monosaccharides, $V^0_{\Phi,S}$, have been evaluated and used to calculate the corresponding transfer volumes, $\Delta_t V^0_{\Phi,S}$, from water to aqueous serine solutions. It is shown that values of transfer volumes and viscosity *B*-coefficients are positive and increase with increasing serine contents. Volumetric interaction parameters, v_{SA}, have been obtained from standard transfer volumes of the monosaccharides. Both values of v_{SA} and viscosity *B*-coefficients are in the order: glucose > galactose; xylose > ribose, which is ascribed to the difference in stereochemistry between *e*–OH and *a*–OH groups. The interactions between monosaccharides and serine are discussed in terms of the structural interaction model and the stereo-structure of monosaccharide molecules. On the other hand, the infrared spectra of the four monosaccharides and 3-O-methyl-D-glycopyranose in deuterium oxide with and without serine were recorded at room temperature. The results confirm further the conclusions from volumetry and viscosity.

---

* Corresponding author. E-mail: klzhuo@263.net

1. INTRODUCTION

Saccharides (S) are important informational molecules, and are the continuation of gene information. Glycomics is of considerable interest more and more. Carbohydrates with proteins or nucleic acids linked through covalent bonds form complex carbohydrates. Specific interactions between carbohydrates and proteins occur in a wide variety of important biological processes, including cell-cell communication, cell adhesion, fertilization, differentiation, development, inflammation, and tumor cell metastasis [1]. Therefore, the understanding of the molecular basis for carbohydrate-protein interactions not only provides valuable information on biological processes in living organisms but it also aids the development of potent biomedical agents [2]. However, the direct thermodynamic studies of the interactions between carbohydrates and proteins in solution are difficult due to their complicated structures. As monosaccharides and amino acids (A) are the most basic units of polysaccharides and proteins, respectively, it is necessary to study the thermodynamic properties for monosaccharide–amino acid–water systems. Thermodynamic properties for these systems are very important not only in investigations of the nature of interactions and the effect of saccharides on the stabilization mechanism of proteins, but also in providing information of stereo-structure of monosaccharide molecules and the possible interaction modes in the structures of carbohydrate-protein complexes [3]. Moreover, they are taken as the basic data for physiological process, medical study and engineering design. Although there are some measurements of volumetric data [4,5], the viscosity studies and especially the IR spectroscopy [6] are rare reported for these systems.

In our previous work [7], volumetric and viscosity properties of aqueous monosaccharide solutions were reported with and without amino acids (glycine and L-alanine) at 298.15 K. In this paper, we explore the interactions between saccharides and L-serine (Ser) in water using infrared spectroscopy besides volumetric and viscosity properties. From these data the standard transfer volumes $\Delta_t V^0_{\Phi,S}$ and viscosity *B*-coefficients for the monosaccharides are calculated. Results are discussed in terms of the cosphere overlap model and the stereochemistry of saccharide molecules. Furthermore, the changes of the infrared spectra provide significant information on the interaction between saccharides and serine. It is expected that these should provide additional information on the effect of saccharides on the stability of globular proteins.

2. EXPERIMENTAL SECTION

2.1. Chemicals

High purity D-(+)-glucose, D-(+)-galactose, D-(+)-xylose, D-(–)-ribose, and 3-O-methyl-D-glycopyranose were purchased from Sigma Chemical Company and L-serine was purchased from Fluka Chemical Company. They were dried under vacuum to constant weight and then stored over P_2O_5 in desiccators. The deionized water (W) was doubly distilled over $KMnO_4$ and has a conductivity of 1.2×10^{-6} s·cm^{-1} at room temperature. The samples in the infrared spectra method were prepared with deuterium oxide (99.9 atom % D).

2.2. Measurement of Densities, Viscosities and Infrared Spectra

Solution densities were measured using a vibrating-tube digital densimeter (model DMA 60/602 Anton Paar Austria) which has been described elsewhere [8-10]. The temperature around the densimeter cell was controlled by circulating water from a constant-temperature bath (Schott, Germany). A CT-1450 temperature controller and CK-100 ultracryostat were employed to maintain the bath temperature at (298.15 ± 0.005) K. The densimeter was calibrated with pure water (the value of density was taken to be 0.997047 $g \cdot cm^{-3}$ at 298.15 K from Kell's data [11]) and dry air. The uncertainty of molalities of monosaccharides and amino acids is evaluated to be less than ± 0.0001 $mol \cdot kg^{-1}$. The uncertainty in density was estimated to be ± 3×10^{-6} $g \cdot cm^{-3}$.

Solution viscosities were measured by a suspended level Ubbelohde viscometer, which was placed in a water thermostat (Schott, Germany), with a flow time of about 200 s for water at 298.15 K. The temperature of the water thermostat was controlled to be as precise as the density measurements. The viscometer was calibrated at (298.15 and 308.15) K with water. The viscosities for water at different temperatures were taken from the literature [12]. Flow time measurements were performed by a Schott AVS310 photoelectric time unit with a resolution of 0.01 s. The estimated uncertainty of experimental viscosity is ± 0.25 %. Solution viscosity η is given by the following equation:

$$\eta / \rho = Ct - K / t \qquad (1)$$

where C and K are the cell constants, t is the flow time. The details of the experimental procedure are given elsewhere [13].

The infrared spectra at room temperature between 4000 and 750 cm^{-1} were recorded on a Fourier transform infrared spectrometer (Model 40, Bio-Rad, America). Demountable cells with barium fluoride windows were used to record the spectra in deuterium oxide solution. A mean of 16 scans with 4 cm^{-1} spectral resolution were accumulated for each case.

3. Results and Discussion

3.1. Apparent Molar Volume

Apparent molar volumes of monosaccharides $V_{\Phi,S}$ were calculated from the equation [14]:

$$V_{\Phi,S} = \frac{M_S}{\rho} - \frac{(1000 + m_{Ser} M_{Ser})(\rho - \rho_{Ser})}{m_S \rho \rho_{Ser}} \qquad (2)$$

where M_S and M_{Ser} are the molar masses of saccharide and serine, m_S and m_{Ser} are the molalities of saccharide and serine, ρ and ρ_{Ser} are the densities of saccharide + serine + water and serine + water solutions, respectively. The uncertainty of $V_{\Phi,S}$ is ± 0.06 at the lowest molality and ± 0.01 $cm^3 \cdot mol^{-1}$ at the highest molality. It has been observed that plots of $V_{\Phi,S}$ versus m_S are well linear. Therefore, infinite–dilution apparent molar volumes $V^0_{\Phi,S}$, which are equal to the standard partial molar volumes, were obtained by least-squares weighted fitting experimental data to the following equation [15,16]:

$$V_{\Phi,S} = V^0_{\Phi,S} + S_V m_S \tag{3}$$

where S_V is the experimental slope. The weighing factor of $V_{\Phi,S}$ used in the regression analysis is $(\delta V_{\Phi,S})^{-2}$, where $\delta V_{\Phi,S} = -(M + 1000/m)\,\delta\rho/\rho^2$, $\delta\rho$ is uncertainty of the solution density and is taken as a constant [13,14], and M is the molar mass of the solute. The resulting values of $V^0_{\Phi,S}$, $V^0_{\Phi,Ser}$ and S_V for saccharides and serine are summarized in Tables 1 and 5, respectively. They are in good agreement with the literature values [17]. The partial molar volumes of transfer of monosaccharides from water to aqueous serine solutions at infinite dilution $\Delta_t V^0_{\Phi,S}$ were evaluated from

Table 1. Infinite-dilution apparent volumes ($V_{\Phi,S}{}^0$ / $cm^3 \cdot mol^{-1}$) and experimental slopes ($S_v{}^*$ /$cm^3 \cdot kg \cdot mol^{-2}$ in parentheses) for saccharides in serine solutions at 298.15 K

$m_{Ser}{}^a$ / $mol \cdot kg^{-1}$	$V^0_{\Phi,S}$ / $cm^3 \cdot mol^{-1}$	S_V / $cm^3 \cdot kg \cdot mol^{-2}$	m_{Ser} / $mol \cdot kg^{-1}$	$V^0_{\Phi,S}$ / $cm^3 \cdot mol^{-1}$	S_V / $cm^3 \cdot kg \cdot mol^{-2}$
	Glucose			Galactose	
0	111.81±0.01	0.69±0.01	0	110.21±0.02	0.57±0.03
	111.99±0.01[b]			110.52±0.02[b]	
	111.79[c]			110.29±0.04[d]	
0.09999	112.09±0.01	0.59±0.01	0.1000	110.37±0.02	0.58±0.02
0.1500	112.19±0.02	0.53±0.02	0.1500	110.46±0.02	0.54±0.03
0.2000	112.25±0.01	0.52±0.02	0.2000	110.56±0.02	0.52±0.02
0.2499	112.46±0.01	0.47±0.01	0.2500	110.61±0.01	0.55±0.01
0.3000	112.48±0.01	0.48±0.01	0.3000	110.65±0.01	0.58±0.01
0.3499	112.53±0.004	0.49±0.01	0.3500	110.73±0.02	0.54±0.02
	Xylose			Ribose	
0	95.47±0.01	0.41±0.01	0	95.22±0.01	0.38±0.01
	95.4±0.3[d]			95.26[f]	
	95.60[e]				
0.1000	95.69±0.01	0.33±0.01	0.1000	95.36±0.004	0.33±0.005
0.1500	95.71±0.004	0.36±0.01	0.1500	95.40±0.01	0.32±0.01
0.2000	95.77±0.01	0.35±0.01	0.2000	95.45±0.01	0.33±0.01
0.2500	95.82±0.01	0.35±0.01	0.2500	95.54±0.01	0.31±0.01
0.3000	95.88±0.01	0.36±0.01	0.3000	95.57±0.01	0.33±0.01
0.3500	95.92±0.01	0.36±0.01	0.3500	95.61±0.01	0.34±0.01

[a] m_{Ser} is the molality of serine in water. [b] Ref. 8. [c] Ref. 17a. [d] Ref. 17b. [e] Ref. 17c. [f] Ref. 17d.

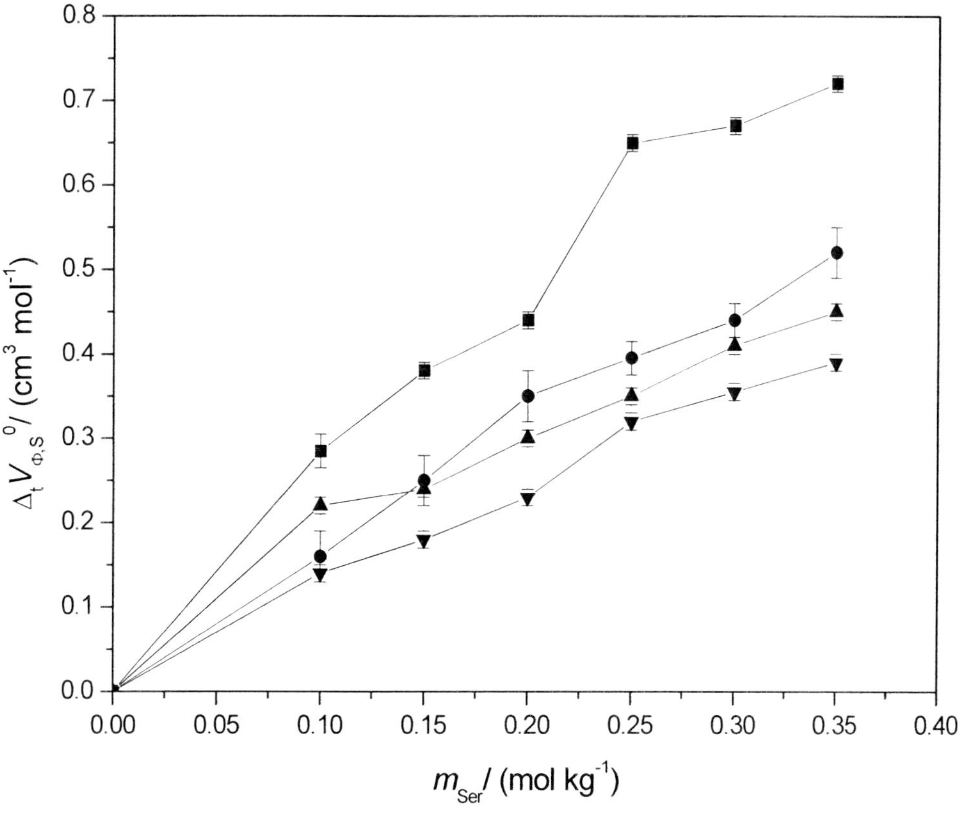

Figure 1. The standard molar volume of transfer ($\Delta_t V^0_{\Phi,S}$) of different saccharides from water to aqueous serine solutions as a function of molality of serine: ■, glucose; ●, galactose; ▲, xylose; ▼, ribose. Vertical bars indicate the standard deviations of the standard molar volume of transfer, which are calculated from the standard deviations of $V^0_{\Phi,S}$ in glycine + water and in water by $\sigma(\Delta_t V^0_{\Phi,S}) = [\sigma(V^0_{\Phi,S}\text{ in glycine + water})^2 + \sigma(V^0_{\Phi,S}\text{ in water})^2]^{1/2}$.

$$\Delta_t V^0_{\Phi,S} = V^0_{\Phi,S}(\text{in aqueous serine}) \quad V^0_{\Phi,S}(\text{in water}) \tag{4}$$

Plots of $\Delta_t V^0_{\Phi,S}$ versus the molalities of serine are represented in Figure 1. Since $V^0_{\Phi,S}$ is, by definition, free from solute-solute interactions, it provides some information about solute–cosolute interactions.

3.1.1. Volumetric Properties of Different Monosaccharides

In aqueous serine, the differences in transfer volumes for different saccharides are ascribed mainly to those in the stereo-structure of saccharide molecules, whose dominant conformations in water are illustrated in Table 2.

Table 2. The dominant conformations of monosaccharides in water

Carbohydrates	Dominant Conformation[a]
D-Glucose	1*e*2*e*3*e*4*e*6*exo*
D-Galactose	1*e*2*e*3*e*4*a*6*exo*
D-Xylose	1*e*2*e*3*e*4*e*
D-Ribose	1*e*2*e*3*a*4*e*

[a] Axial (*a*) , equatorial (*e*) or exocyclic (*exo*) hydroxyl groups.

First, it is seen from Figure 1 that the $\Delta_t V^0_{\Phi,S}$ values for all saccharides studied are positive and increase with increasing molality of serine. This can be rationalized by the cosphere overlap model, as developed by Friedman and Krishnan [18]. The interactions between the monosaccharide and amino acid molecules can be classified as follows: (1) Hydrophilic–ionic interactions between the OH groups of monosaccharide and the zwitterionic center of amino acid; (2) Hydrophilic–hydrophilic interactions between the OH groups of monosaccharide and the OH groups of amino acid mediated through the hydrogen bond; (3) Hydrophilic–hydrophobic interactions between the OH groups of monosaccharide/amino acid and the non-polar groups of the amino acid /monosaccharide; (4) Hydrophobic–hydrophobic interactions between the non-polar groups of monosaccharide and amino acid. The interations of types (1) and (2) make positive contributions to the transfer volume, whereas the contributions of types (3) and (4) are opposite. Therefore, the observed increasing positive transfer volumes suggest that the first two interactions are predominant in the ternary systems studied. In addition, an increase in concentration of amino acid leads to stronger interactions of types (1) and (2), consequently the $\Delta_t V^0_{\Phi,S}$ values of saccharides increase.

Secondly, as presented in Figure 1, the $\Delta_t V^0_{\Phi,S}$ values for saccharides in aqueous serine solution are in the order: glucose > galactose > xylose > ribose except at low molality of serine (m_{Ser} = 0.15 mol·kg^{-1}). For pentoses or hexoses, all structural isomers are essentially different only in the stereochemical structure of hydroxyl group on the ring (equatorial, axial or exocyclic position). The experimental results, which are concordant with our previous work [5], give an evidence about the following supposition. The interaction of *e*–OH with amino acids contributes more positive value to $\Delta_t V^0_{\Phi,S}$ than *a*-OH group. Therefore, the values of $\Delta_t V^0_{\Phi,S}$ for galactose and ribose are smaller than those for glucose and xylose, respectively. On the other hand, the positive contribution of *e*–OH to $\Delta_t V^0_{\Phi,S}$ is larger than the sum of *a*–OH and *exo*–OH at low molality of serine so that there are smaller values for galactose than those for xylose at m_{Ser} = 0.15 mol·kg^{-1} (see Figure 1). But at high concentration of seine (m_{Ser} > 0.15 mol·kg^{-1}), the *a*–OH and *exo*–OH play an important role and lead to the order of $\Delta_t V^0_{\Phi,S}$ values: galactose > xylose.

McMillan and Mayer [19] proposed a solution theory that permits the formal separation of the interactions between two or more solute molecules. According to this treatment, at infinite dilution of amino acid (A) for the S–A–W system, $\Delta_t V^0_{\Phi,S}$ can be expressed as:

Table 3. Volumetric pair and triplet interaction parameters for saccharide–serine–water solutions at 298.15 K

Saccharide	ν_{SA} /cm^3·mol^{-2}·kg	ν_{SAA} /cm^3·mol^{-2}·kg^2	R
Glucose	1.484±0.153	-0.826±0.353	0.977
Galactose	0.927±0.059	-0.369±0.135	0.994
Xylose	0.988±0.089	-0.683±0.204	0.970
Ribose	0.684±0.055	-0.223±0.126	0.991

$$\Delta_t V^0_{\Phi,S} = 2\nu_{SA} m_A + 3\nu_{SAA} m^2_{AA} + \ldots \qquad (5)$$

where ν_{SA} and ν_{SAA} are pair and triplet interaction parameters, respectively. These parameters were obtained from the fit of experimental data to the equation and given in Table 3, together with correlation coefficients. It appears from Table 3 that ν_{SAA} lacks precision because of experimental errors. However, this is not too important because we are interested principally in the pair interaction parameter ν_{SA}, which is the most important in dilute solutions. The values of ν_{SA} follow the order: glucose > xylose > galactose > ribose, which may be due to the difference in stereo-structure of monosaccharides (see Table 2). This supports further the conclusion that at very dilute solutions of amino acids, the contribution of *e*–OH to $\Delta_t V^0_{\Phi,S}$ is predominant over the sum of *a*–OH and *exo*–OH.

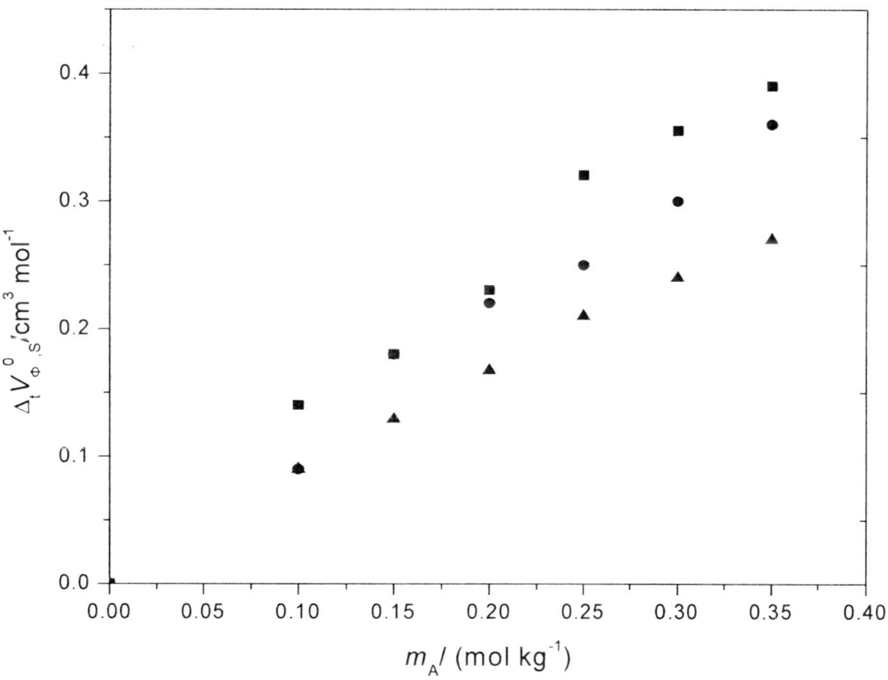

Figure 2. Standard volumes of transfer of ribose from water to aqueous amino acid solutions as a function of molality of amino acid at 298.15 K: ■, ribose in serine; ●, ribose in glycine; ▲, ribose in alanine. Vertical bars indicate the standard deviations of $\Delta_t V^0_{\Phi,S}$ (see the caption of Figure 1).

3.1.2. Volumetric Properties of Different Amino Acids

Combined with data reported in our previous work [7], variations of the transfer volumes of ribose to different amino acid solutions are represented in Figure 2. It can be seen from this figure that the $\Delta_t V^0_{\Phi,S}$ values for each of four saccharides considered decrease overall at given m_A in the same order: serine > glycine > alanine. Serine is regarded as a derivate of glycine whose one hydrogen atom of the α-carbon is replaced by a CH_2OH group, and consequently the hydrophility of serine is stronger than glycine and alanine, which leads to more positive values of $\Delta_t V^0_{\Phi,S}$. As a result, the $\Delta_t V^0_{\Phi,S}$ values for monosaccharides from water to serine+water are more positive than to glycine + water or alanine + water.

3.2. Viscosity *B*-coefficient

The relative viscosity η_r can be analyzed using the Jones-Dole equation [20]:

$$\eta_r = \eta / \eta_0 = 1 + Bc + Dc^2 \tag{6}$$

where η is the viscosity of the ternary solutions (S + Ser + W), η_0 is the viscosity of the binary solutions (Ser + W), c (mol·dm^{-3}) is the molarity of saccharides, and B and D are empirical coefficients known as the viscosity *B*- and *D*-coefficients. Values of B and D were obtained from a linear plot of $(\eta_r - 1)/c$ versus c. Viscosity *B*-coefficients are collected in Table 4. Recent literature data [21] for glucose in water are also represented in Tables 4 and 5. The *B*-coefficients measure the size and shape effects as well as the structural effect induced by solute-solvent interactions [22,23], whereas *D*-coefficients reflect solute-solute interactions as well as solute-solvent interactions. Therefore, *D*-coefficients are too complex to analyze especially in the presence of serine and are not considered here.

Table 4. Viscosity *B*-coefficients of saccharides in aqueous serine solution at 298.15 K

m_{Ser} / mol·kg^{-1}	B / dm^3·mol^{-1}	m_{Ser} / mol·kg^{-1}	B / dm^3·mol^{-1}	m_{Ser} / mol·kg^{-1}	B / dm^3·mol^{-1}	m_{Ser} / mol·kg^{-1}	B / dm^3·mol^{-1}
Glucose		Galactose		Xylose		Ribose	
0	0.440±0.004 0.45[a]	0	0.416±0.004	0	0.336±0.001	0	0.319±0.004
0.1000	0.441±0.003	0.1000	0.423±0.003	0.1000	0.353±0.001	0.1000	0.303±0.003
0.1500	0.442±0.003	0.1500	0.427±0.003	0.1500	0.356±0.001	0.1500	0.306±0.002
0.2000	0.443±0.003	0.2000	0.432±0.004	0.2000	0.358±0.001	0.2000	0.309±0.002
0.2500	0.447±0.003	0.2500	0.438±0.004	0.2500	0.359±0.001	0.2500	0.319±0.001
0.3000	0.448±0.003	0.3000	0.442±0.003	0.3000	0.362±0.002	0.3000	0.322±0.002
0.3500	0.451±0.001	0.3500	0.445±0.003	0.3500	0.365±0.001	0.3500	0.325±0.002

[a] Ref. 21.

Table 5. Solution densities ρ, infinite–dilution apparent volumes ($V_{\varphi,s}^{0}$ / $cm^3 \cdot mol^{-1}$) and viscosity B-coefficients for serine in water at 298.15 K

m_{Ser}[a] / $mol \cdot kg^{-1}$	ρ_{Ser} / $g \cdot cm^{-3}$	$V_{\Phi,Ser}$ / $cm^3 \cdot mol^{-1}$	$V^0_{\Phi,Ser}$ / $cm^3 \cdot mol^{-1}$	B / $dm^3 \cdot mol^{-1}$
		Serine		
0.1000	1.001477	60.57	60.39±0.003[b]	0.242±0.002
0.1500	1.003658	60.66	60.23±0.05[c]	0.238[f]
0.2000	1.005818	60.75	60.26±0.02[d]	0.207[c]
0.2500	1.007955	60.84	60.62[e]	
0.3000	1.010072	60.93		
0.3500	1.012166	61.02		

[a] m_{Ser} is the molality of serine in water. [b] the experimental slope S_v^* of serine is 1.81 ± 0.01 $cm^3 \cdot kg \cdot mol^{-2}$. [c] Ref. 17e. [d] Ref. 4. [e] Ref. 17f. [f] Ref. 17g.

It can be seen from Table 4 that all the viscosity B-coefficients for saccharides are positive and decrease both in water and in aqueous serine solutions in the order: glucose > galactose > xylose > ribose. Glucose and galactose have similar size and shape, similarly for xylose and ribose, and consequently this order of viscosity B-coefficients may be explained in terms of the solute-solvent interactions. Compared with galactose or ribose, glucose or xylose interacts more strongly with serine. It can be concluded that the *e*–OH interacts with serine more strongly than *a*–OH. The stronger interaction offers the larger resistance to the movement of solute particles (monosaccharides). Therefore, there is the order: B(glucose) > B(galactose), B(xylose) > B(ribose). Similarly, we can explain why the viscosity B-coefficients increases with increasing the amino acid content in the ternary system. In addition, hexose has larger volume than pentose, and the size effect of solute molecules results in larger viscosity B-coefficients of hexose than pentose.

3.3 Infrared Spectra

In recent years a number of studies on the calculation of frequencies and the forms of normal vibrations of mono- and disaccharides have appeared [24-27]. And these regions are generally adopted for their significance in the structural analysis of saccharides: (a) 3650 ~ 3200 cm^{-1}: OH stretching vibrations; (b) 3000 ~ 2800 cm^{-1}: CH stretching vibration; (c) 1500 ~ 1128 cm^{-1}: HCH, CH_2OH vibration; (d) 1120 ~ 1000 cm^{-1}: C–O–C stretching region. The infrared spectra of various saccharides at the same molality (0.6 $mol \cdot dm^{-3}$) and the saccharides–serine–D_2O system were recorded. Fourier transform infrared spectra are collected to observe the characteristic spectrum pattern of different saccharides and provide some information of the interactions between saccharide and serine.

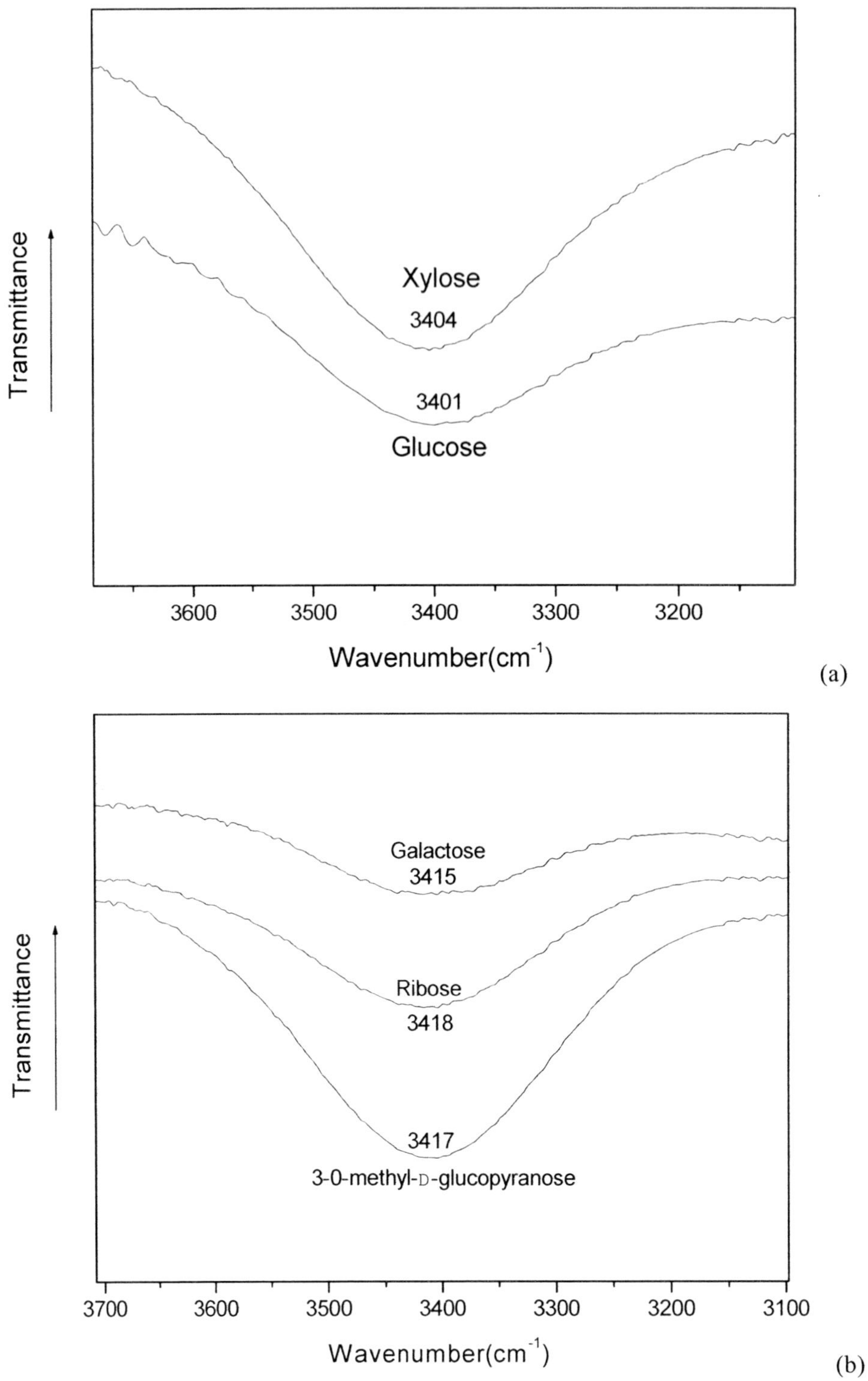

Figure 3. Infrared absorption spectra in the range of 3650 ~3200cm^{-1} of (a) glucose and xylose; (b) galactose, ribose, and 3-O-methyl-D-glucopyranose.

3.3.1. The Characteristic Spectrum Pattern of Different Saccharides

(1) The region 3650~3200 cm^{-1}. It should be noticed that hydrogen bonds can be formed between saccharide molecules in solution, so the vibrational frequency of OH shifts to the lower position. From Figures 3(a) and (b), we can see that the vibrational frequencies of glucose are similar to xylose, and galactose to ribose respectively, moreover, glucose < galactose, xylose < ribose. And the infrared intensity is in the order: glucose > galactose, xylose > ribose. This shows that *e*–OH is easier to form hydrogen bond compared to *a*-OH. Consequently, the vibrational frequency for 3-O-methyl-D-glucopyranose shifts to higher frequency compared with glucose because its one OH group is replaced by the OCH_3 group to form 3-O-methyl-D-glucopyranose.

(2) The region 1120 ~ 1000 cm^{-1}: Because of the common oxygen atom in the group C–O–C on the ring of saccharides, strong coupling occurrs. Figure 4 shows that the difference between these two absorption peaks concerning infrared intensities of glucose and xylose are stronger than 3-O-methyl-D-glucopyranose, galactose and ribose. Therefore, the *e*–OH group contributes positively to the difference of the two split peaks' absorption intensities. It is concluded that vibrations have a peculiar localization and that steric factors play an important role in the vibrational spectra of saccharides.

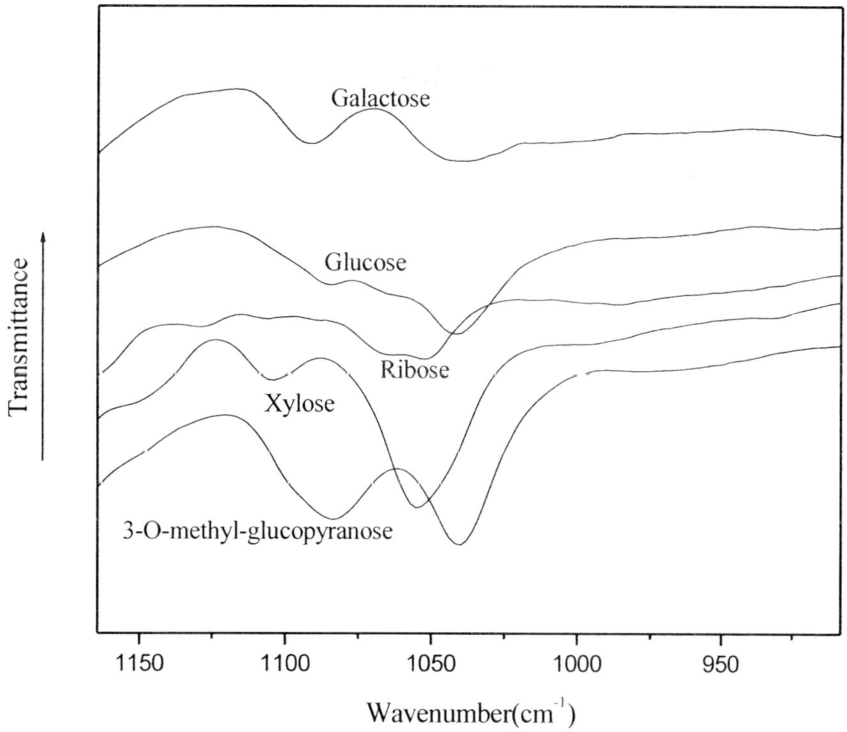

Figure 4. Infrared absorption spectra in the range of 1000~1120 cm^{-1} of glucose, xylose, galactose, ribose, and 3-O-methyl-D-glucopyranose.

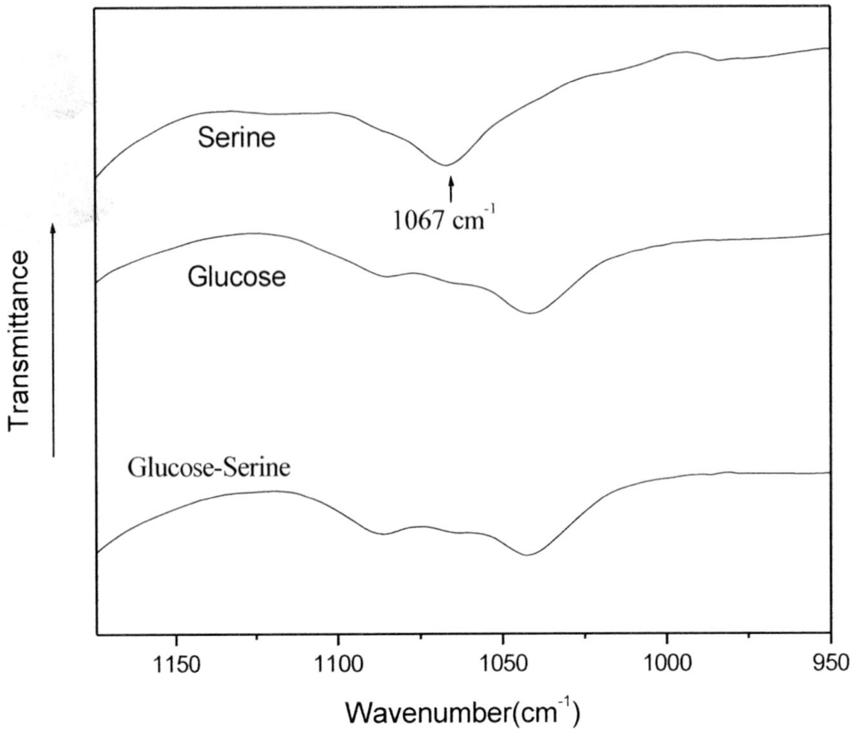

Figure 5. Infrared absorption spectra in the range of 1000~1120 cm^{-1} for glucose in D_2O (0.6 mol·dm^{-3}), serine in D_2O (0.25 mol·dm^{-3}) and glucose–serine in D_2O (m_S = 0.6 mol·dm^{-3}, m_{Ser} = 0.25 mol·dm^{-3}).

3.3.2. The Comparison of Infrared Spectra of Serine in D_2O and Serine–saccharides in D_2O

To determine the interaction between saccharides and serine, a comparative study was made for serine in D_2O and serine–saccharides in D_2O where the molarities of serine are the same (0.25 mol·dm^{-3}). The infrared spectrum of glucose–serine in D_2O was taken as an example (Figure 5), it is observed that the infrared spectrum of glucose–serine in D_2O is similar to glucose in D_2O, and the peak at 1067 cm^{-1} in the spectrum of serine in D_2O has disappeared indicating that the vibration of the C–O bond in serine is influenced by saccharide molecules. We presume that this oxygen atom has a relatively strong interaction with the OH groups of saccharides, which is important evidence about the types of interaction between saccharides and serine.

4. CONCLUSION

We have investigated the interactions of saccharides with serine in water. The results show that the hydrophilic–zwitterionic and hydrophilic–hydrophilic interactions are predominant in saccharide–serine–water systems studied. Both values of ν_{SA} and viscosity *B*-coefficients are in the order: glucose > galactose; xylose > ribose, which is ascribed to the difference between *e*–OH and *a*–OH groups, namely, the *e*–OH interacts with serine more strongly and forms hydrogen bond more easily than *a*–OH. This confirms that stereochemistry of saccharides play an important role in their hydration and interactions with other co-solutes, especially large molecules in size. The study of FT-IR spectroscopy supports further these conclusions. On the other hand, it shows also that the OH group linked to CH_2 in serine molecule has a relatively strong interaction with the OH groups of saccharides.

ACKNOWLEDGMENT

This work was financially supported by the National Natural Science Foundation of China (grant 20673033) and the Innovator Foundation of Colleges and Universities of Henan Province.

REFERENCES

[1] Roth, J. *Chem. Rev.* 2002, *102*, 285-304. (b) Ritchie, G. E.; Moffatt, B. E.; Sim, R. B.; Morgan, B. P.; Dwek, R. A.; Rudd, P. M. *Chem. Rev.* 2002, *102*, 305-320. (c) Zachara, N. E.; Hart, G. W. *Chem. Rev.* 2002, *102*, 431-438. (d) Bertozzi, C. R.; Kiessling, L. L. *Science* 2001, *291*, 2357-2364.

[2] Danishefsky, S. J.; Allen, J. R. A. *Angew. Chem., Int. Ed.* 2000, *39*, 836-863. (b) Walter, F.; Vicens, Q.; Westhof, E. *Curr. Opin. Chem. Biol.* 1999, *3*, 694-704. (c) Sears, P.; Wong, C. H. *Angew. Chem., Int. Ed.* 1999, *38*, 2301-2324.

[3] Mazik, M.; Cavga, H; Jones, P. G. *J. Am. Chem. Soc.* 2005, *127*, 9045-9052.

[4] Li, S. Q.; Sang, W. Q.; Lin, R. S. *J. Chem. Thermodyn.* 2002, *34*, 1761-1768.

[5] Ali, A.; Hyder, S.; Sabir, S.; Chand, D. A.; Nain, K. *J. Chem. Thermodyn.* 2006, *38*, 136-143.

[6] Mathlouthi, M. J.; Koenig, J. L. *Adv. Carbohydr. Chem. Biochem.* 1986, *44*, 7-89.

[7] Zhuo K. L.; Liu, Q.; Wang, Y. P.; Ren, Q. H.; Wang, J. J. *J. Chem. Eng. Data* 2006, *51*, 919-927.

[8] Zhuo, K. L.; Wang, J. J.; Yue, Y. K.; Wang, H. Q. *Carbohydr. Res.* 2000, *328*, 383-391.

[9] Zhuo, K. L.; Wang, J. J.; Wang, H. Q.; Yue, Y. K. *Z. Phys. Chem.* 2001, *215*, 561-573.

[10] Zhuo, K. L.; Wang, J. J.; Zheng, H. H.; Xuan, X. P.; Zhao, Y. *J. Solution Chem.* 2005, *34*, 155-170.

[11] Kell, G. S. *J. Chem. Eng. Data* 1967, *12*, 66-69

[12] James, C. J.; Mulcahy, D. E.; Steel, B. J. *J. Phys. D: Appl. Phys.* 1984, *17*, 225-230.

[13] Yan, Z. N.; Wang, J. J.; and Lu, J. S. *Biophys. Chem.* 2002, *99*, 199-207.

[14] Visser, C. D.; Perron, G.; Desnoyers, J. E. *J. Am. Chem. Soc.* 1977, *99*, 5894-5900.

[15] Hedwig, G. R. *J. Solution Chem.* 1988, *17*, 383-397.

[16] Zielenkiewicz, W.; Perlovich, G. L.; Nikitina, G. E.; Semeykin, A. S. *J. Solution Chem.* 1996, *25*, 135-153.

[17] (a) Jasra, R. V. and Ahluwalia, J. C. *J. Chem. Thermodyn.* 1984, *16*, 583-590. (b) Hoiland, H.; Holvik, H. *J. Solution Chem.* 1978, *7*, 587-596. (c) Jasra, R. V.; Ahluwalia, J. C. *J. Solution Chem.* 1982, *11*, 325-338. (d) Banipal, P. K.; Banipal, T. S.; Ahluwalia, J. C.; Lark, B. S. *J. Chem. Thermodyn.* 2002, *34*, 1825-1846. (e)

Belibağlı, K. B.; Ayrancı, E. *J. Solution Chem.* 1990, *19*, 867-882. (f) Jolicoeur, C.; Riedl, B.; Desrochers, D.; Lemelin, L. L.; Zamojka, R.; Enea, O. *J. Solution Chem.* 1986, *15*, 109-128. (g) Yan, Z. N.; Wang, J. J.; Liu, W. B.; Lu, J. S. *Thermochim. Acta* 1999, *334*, 17-27.

[18] Friedman, H. L.; Krishnan, C. V. *In Water: A Comprehensive Treatise*. Franks, F., Ed.; Plenum: New York 1973, Vol. *3*. Chapter 1.

[19] McMillian, W. G.; Mayer, J. *J. Chem. Phys.* 1945, *13*, 276-305.

[20] Jones, G.; Dole, M. *J. Am. Chem. Soc.* 1929, *51*, 2950-2964.

[21] Miyajima, Koichiro; Sawada, Masahiro; Nakagaki, Masayuki. *Bull. Chem. Soc. Jpn.* 1983, *56*, 1954-1957.

[22] Stokes, R. H.; Milles, R. *International Encyclopedia of physical Chemistry and Chemical Physics*; Pergamon: New York 1965.

[23] Nakagawa, T. *J. Mol. Liq.* 1995, *63*, 303-316.

[24] Hineno, M. *Carbohydr. Res.* 1977, *56*, 219-227.

[25] Korolevich, M. V.; Zhbankov, R. G., Sivchik, V. V. *J. Mol. Struct.* 1990, *220*, 301-313.

[26] Yang, L. M.; Weng, S. F.; Ferraro, J. R.; Wu, J. G. *Vibrational Spectroscopy*. 2001, *25*, 57-62.

[27] Zhbankov, R. G.; Andrianov, V. M.; Marchewka, M. K. *J. Mol. Struct.* 1997, *436-437*, 637-654.

INDEX

A

B

S

T

U

V

W

X

Y

Z